GUNBOAT SOEMBA

THE TERRIBLE TWIN SISTER

SOEMBA

A gunboat is a small, armed naval vessel primarily designed for coastal defence and patrol duties. Faster and more manoeuverable than larger warships, allowing them to operate effectively in coastal and inland waters. These ships played a significant role in Dutch naval history, particularly in defending coastal waters and protecting merchant shipping.

INTRODUCTION

In the years following the First World War, the Royal Netherlands Navy in the East Indies was reinforced with several new units. First of all, the light cruisers *Java* and *Sumatra* should be mentioned (the construction of the third cruiser, *Celebes*, did not proceed). In addition, two artillery ships were put into service, armed, like the aforementioned light cruisers, with 15 cm guns. These were the flotilla leaders *Flores* and *Soemba*, which, unlike the cruisers intended for the Dutch squadron, had to serve as local defence.

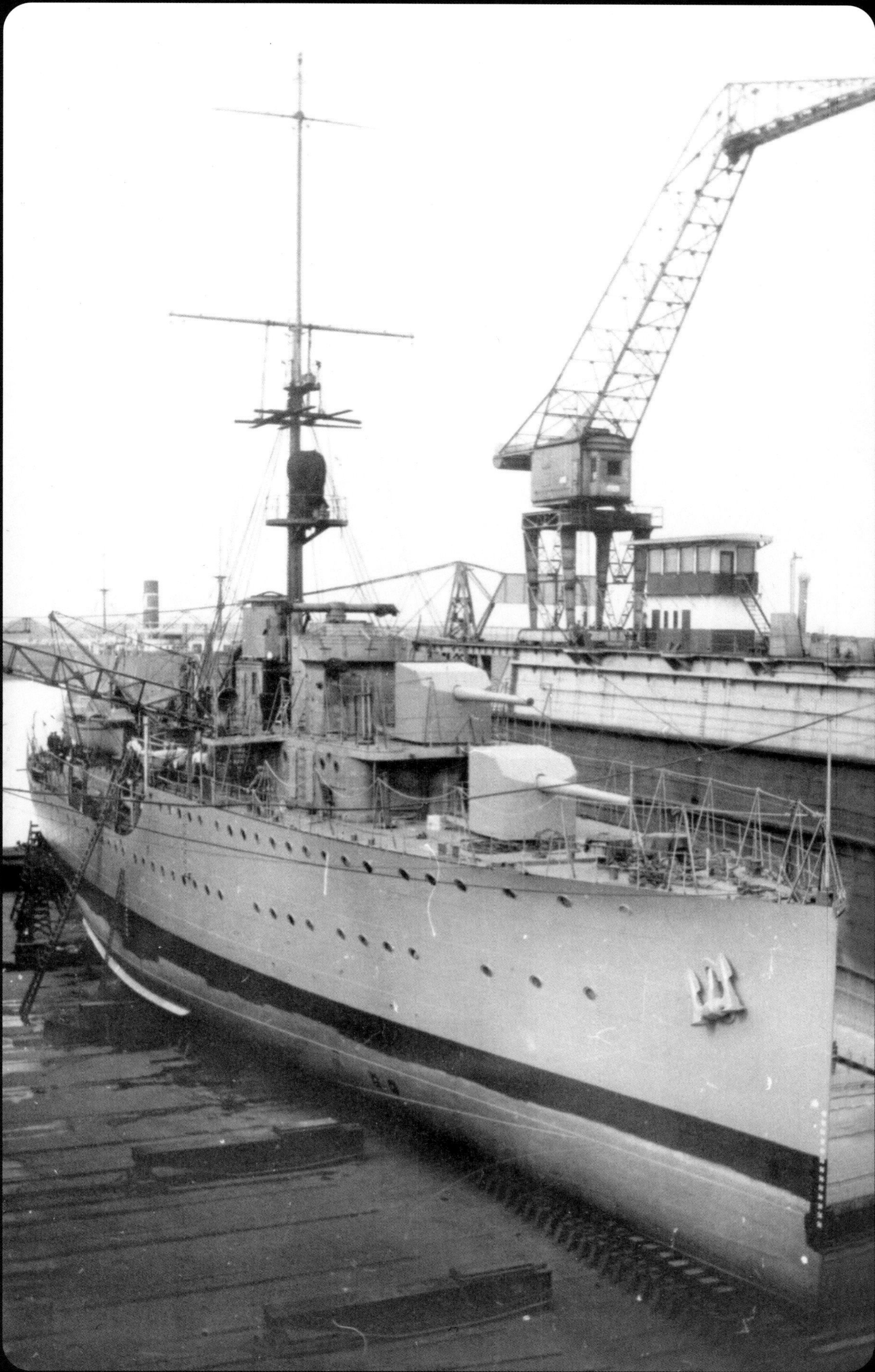

FLORES-CLASS GUNBOATS

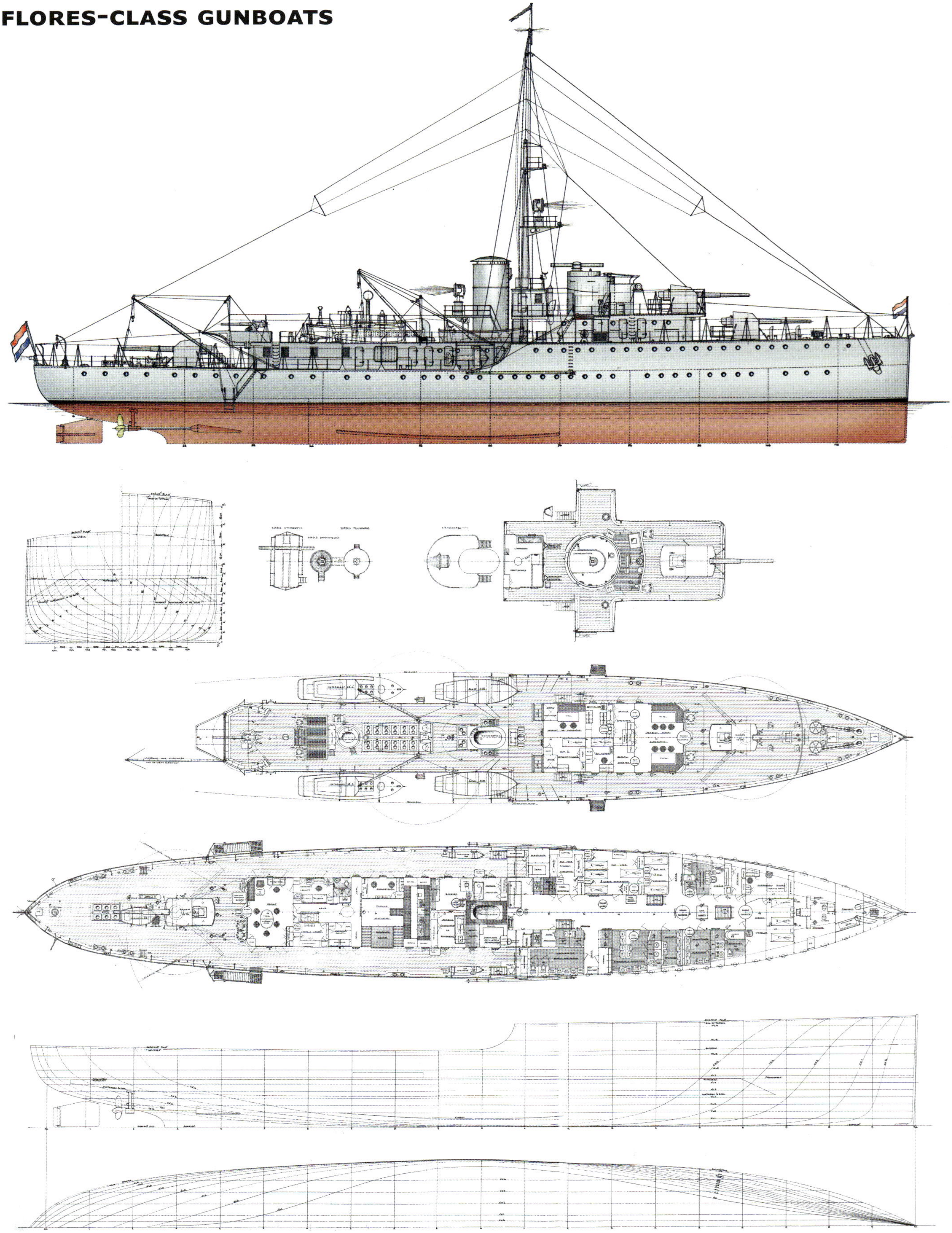

Service in the Netherlands Indies in the 1930s.
(NIMH)

GUNBOATS*

The Royal Netherlands Navy has always called the relatively small vessels used for police surveillance in the East Indian Archipelago "flotilla vessels". They were approximately 500 ton displacement ships, armed with a few 10.5 cm cannons, which were completely sufficient for that purpose. However, over the years they become outdated and completely worn out, so that in the 1920s replacement was urgently necessary. There was also a need for a few armoured boats to be able to defend the mine barriers around Surabaya in times of war. It was obvious to combine the two types a new one to be designed, suitable for both purposes. The obvious solution was to combine the two types into a new dual-purpose design, for which the following requirements were established:

1. As law enforcement vessel:

a. large radius of action,
b. simple, robust machinery,
c. low draft (3.60 m), with a view to navigating rivers,
d. great manoeuvering ability (for the same purpose),
c. spacious accommodations, suitable for long voyages in the tropics.

2. As warship:

a. an armament of 3 guns of 15 cm and 1 gun of 7.5 cm, in order to be able to cope with any vessel used for clearing my obstacles, such as torpedo boats, armed trawlers, etc.,
b. some protection, namely an armoured deck, an armoured conning tower, and armoured ammunition chutes. It is of course not possible to make this armour resistant to direct impact. Therefore, it was exclusively shrapnel-resistant, as was the case on the then new light cruisers
c. a highly advanced watertight layout.

* Taken from magazine 'Het Schip' of 26 August 1926

Pre-WWII photo of Soemba *in Den Helder.*

Also, Flores *was in WWII modified. Several anti aircraft guns and splintershields were added, as can be seen in this photo.*

Armament and armour

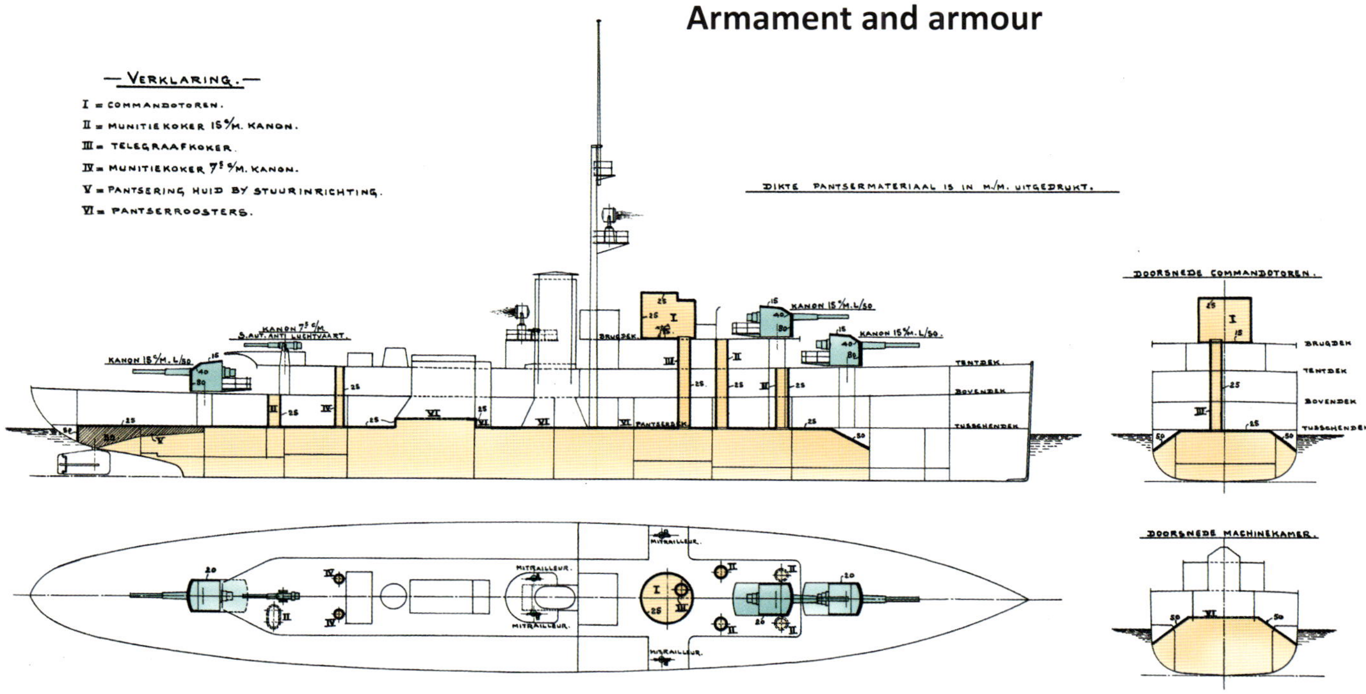

For both services, a speed of 14 miles, which could be maintained under all circumstances, was considered sufficient. In connection with this, the test-run speed was set at 15 miles. It was also determined that the two forward guns would fire over each other, and that the ships would be provided with a forecastle. - this meant that various weights (2 guns, the deckhouse and the conning tower) had to be placed very high, which meant that stability required special attention. On the other hand, a large space became available under the forecastle, which greatly improved the habitability.

It was known in advance that the heavy armament and armour would considerably exceed the main dimensions and displacement of the previous flotilla vessels. The length was to be 69 m according to the original estimate. When making the resistance calculations, however, it appeared possible to achieve a considerable reduction in power, by increasing it, mainly because, since the surface of the main frame could not be chosen arbitrarily due to the limited draft, the greater length made it possible to use a smaller cylinder coefficient.

The width was chosen, after a preliminary calculation of the height of the centre of gravity had been made, so that the metacentre height would amount to approximately 50 and. The depth is the sum of the draft, the height of the armoured deck above water and the deck height, and was therefore fixed without further ado.

Main dimensions	
Length overall	75.60 m
Length on C.W.L.	74.00 m
Width	11.50 m
Depth	6.30 m
Draught	3.60 m
Displacement	1683 tons

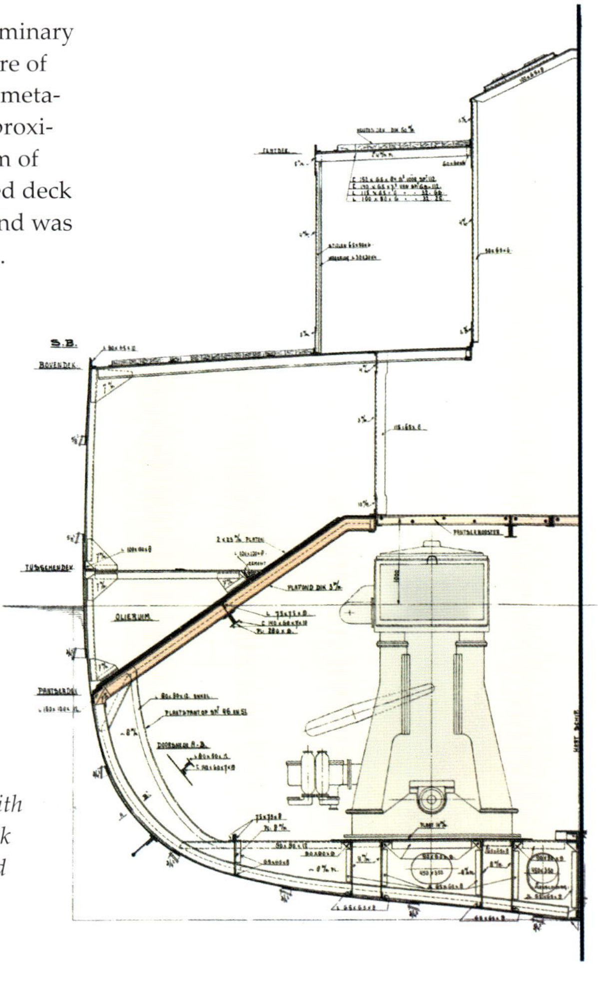

Classical practice to cover the vitals with waterline armour, and to use a flat deck to protect them from plunging hits and debris of superstructure hits.

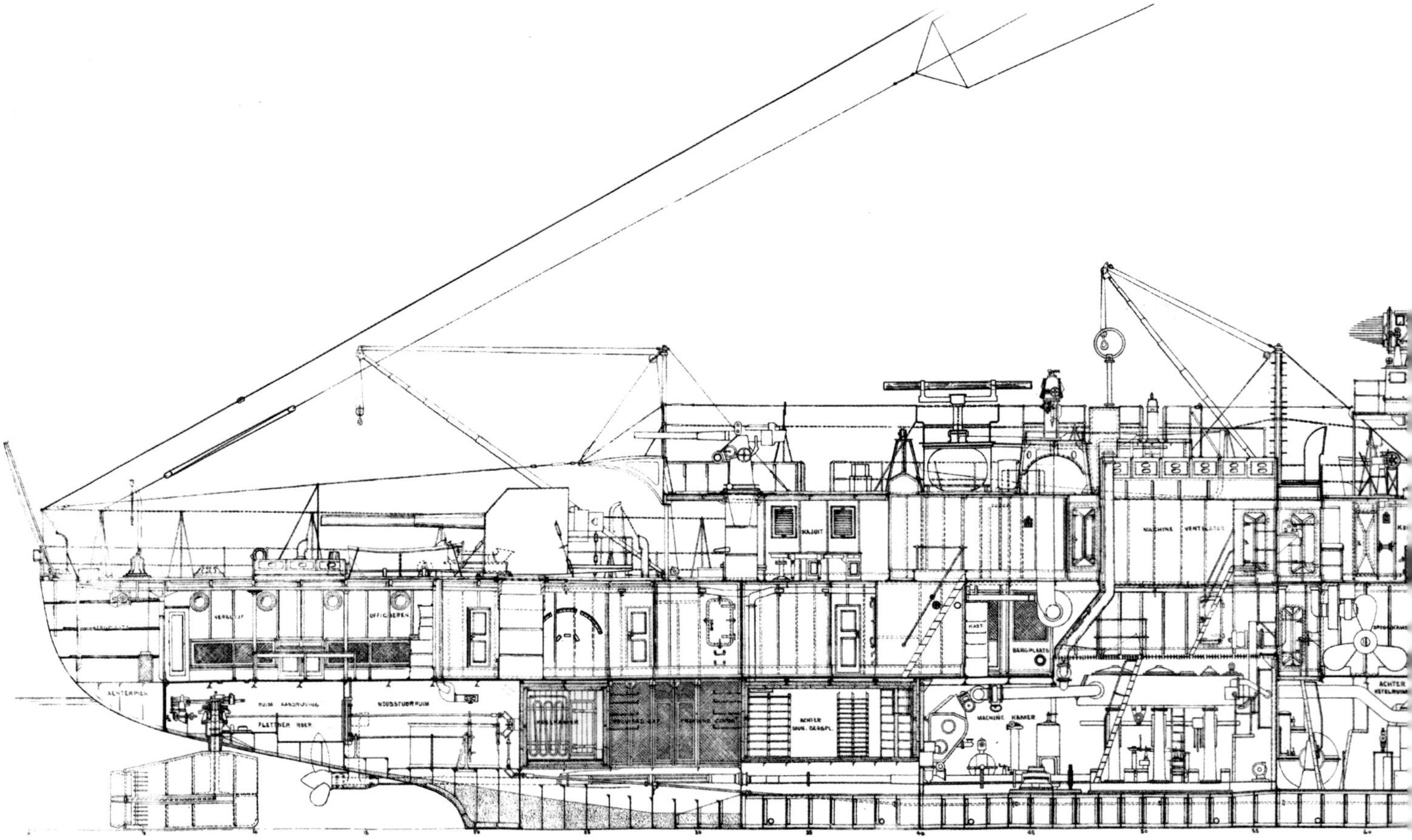

Internal layout

In the lower ship, under the armoured deck, are the engine and boiler rooms. Due to the low speed, these take up relatively little space. Furthermore, here are various storage areas, ammunition rooms, the steering room, etc. The oil was mainly stored in side tanks. This has the advantage that when the tanks are emptied, the influence on stability is small. The large weight in the sides also gives a large mass moment of inertia, and therefore small rolling movements in sea conditions.

On the between deck, in the fore ship, are the quarters for the crew and in the after ship those for the officers. Because the between deck has no sheer, while the upper deck does, these quarters, especially the long room, are very high and airy. Above the engine and boiler rooms, no quarters have been installed with a view to the stay in the tropics; above the boiler rooms are some storage places and the engine workshop; above the engine room are the cooling machine and diesel generator.

On the upper deck (see page 2), under the forecastle in the front, is the provost, the anchor engine room, a spacious sick room with examination room, bath and pharmacy, and, further aft, spacious washrooms for non-commissioned officers and enlisted men, and cabins for non-commissioned officers. Around the funnel are galleys, bakery and bottling room, while the rear part of the deckhouse is occupied by the commander's quarters.

On the tent deck are some cabins for the skipper, and two sergeant majors, spacious quarters for non-commissioned officers and some storage places. On the bridge deck are the steering bridge, the conning tower, the chart room and the guard cabin for the commander.

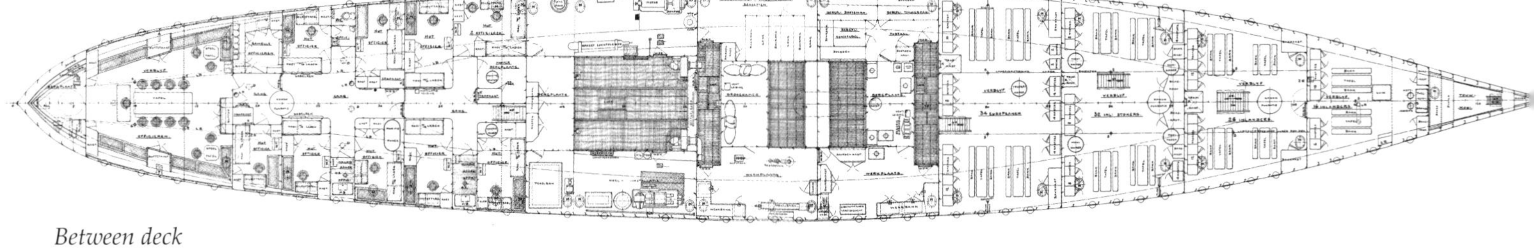

Between deck

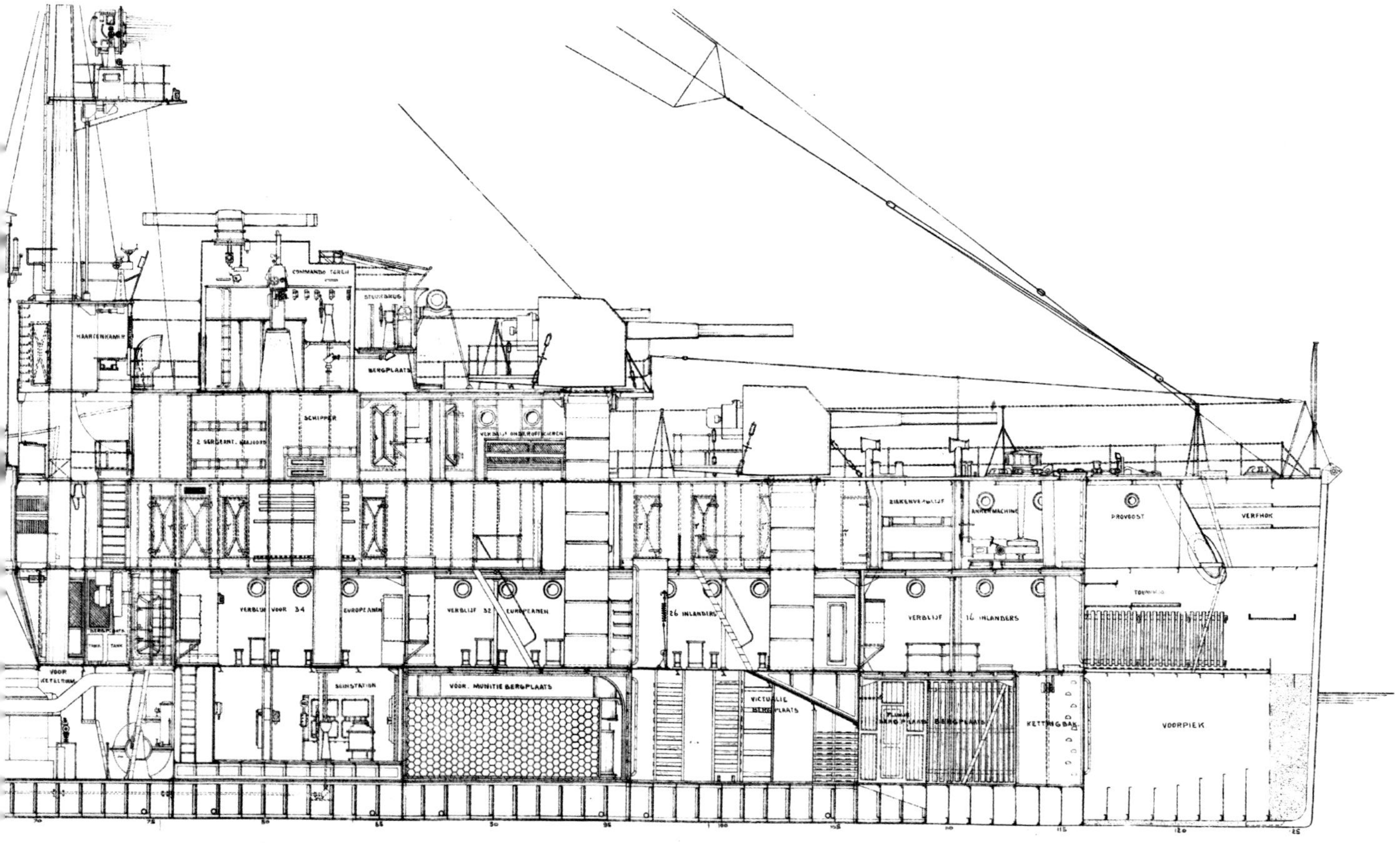

Construction.
In view of the long and heavy service in the tropics it was decided that the skin thickness underwater in the midships would not be less than 9.5 mm. For the skin passages 7 mm was allowed. This resulted in a greater resistance moment of the main frame than the "Board of Trade" prescribes for passenger ships. Special provisions to obtain longitudinal strength were therefore not necessary.

The transverse strength is guaranteed by the numerous watertight bulkheads. Where these are at a great distance, such as in the engine room and in the long room, plate frames are fitted. The bulkheads above the armoured deck, which are very light (3 mm), have a heavier plate along the skin, which serves as a plate frame. The frame profile is under the armoured deck in the engine room reinforced with retaining frames; above the armoured deck, where the frames have no other function have to strengthen the skin and to support the deck. The forward frames are an exception; these have a heavy load due to the beating of the waves when the ship steams up the sea. That is why the frames, for Frame 106, which are no longer broken by the armoured deck, have been extended to the forecastle deck.

The bottom construction is light. A double bottom could not, or only locally, be fitted due to the limited available height under the armoured deck.

The free-standing decks are covered with wood due to service in the Indies. Due to the risk of fire, a full steel deck has been welded underneath everywhere. Teak wood, according to experience, seems to burn poorly when there is steel underneath. In the accommodation, the wood could be omitted as deck covering. Because the beams are placed every other frame, the plating would become too soft. That is why it is reinforced at the bottom with

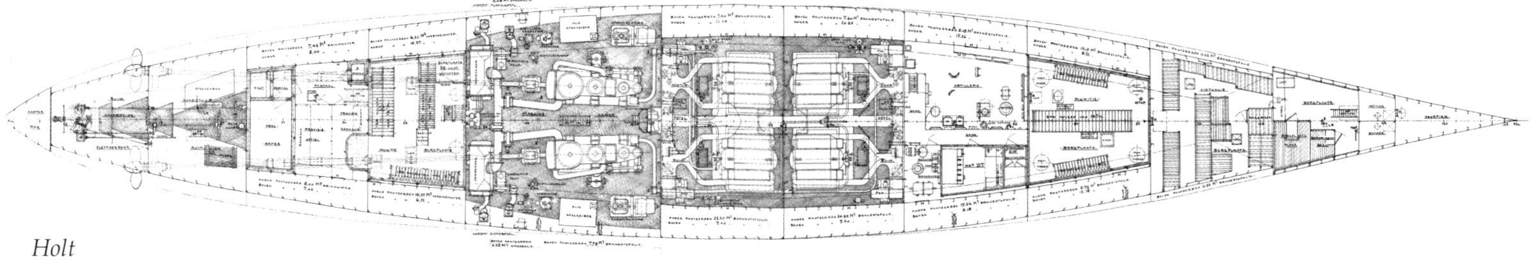

Holt

The Name

Soemba was named after the Indonesian island Sumba (part of the Lesser Sunda Islands) located in the Eastern Indonesia and administratively part of the East Nusa Tenggara provincial territory. Soemba has an area of 11,243.78 square kilometres (4,341.25 square miles), about the same size as Jamaica or Hawaii (Island).:

Name bearers

1. 1926 - 1985
   Flores-class gunboat. The subject of this book
2. 1989 -
   The Soemba was built in 1989 for the army's diving training programme. The training vessel entered naval service on 6 June 2009, when the army and navy merged their diving training programmes. The *Soemba* operates primarily on Dutch inland waterways and is the only green-painted vessel in the Royal Netherlands Navy.

Flettner rudder

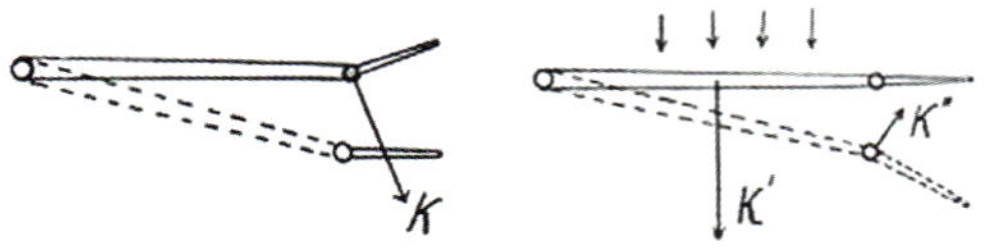

The Flettner Rudder is equipped with a large rudder blade called a flap at the trailing edge of the main rudder blade, which is hinged with a pin between the main and auxiliary rudder. It has similar hydrodynamic characteristics to the flaps on an aircraft. The flaps are connected to the trailing edge of the main rudder blade and can be independently controlled by a separate mechanism. The rotation direction of the flaps can be consistent with the main rudder blade, but the angle is greater than the rudder blade angle. There are also flaps that rotate in the opposite direction to the main rudder blade. Flap rudder can improve rudder efficiency or reduce rudder torque.

Soemba *on patrol the East Indies before WWII. The ship on the left is the auxiliary vessel* Zuiderkruis *of the Dutch Gouvernmental Navy.* (NIMH)

riveted loose pieces of angle iron. This construction method has been in use by the Netherlands navy for many years.

Above the armoured deck the posts are at a distance of 75 cm and the plating is only 3 mm thick, the footplate 4 mm. A plate of 3 mm thickness is fitted in the side, which serves as a plate frame.
The use of wood has been avoided as much as possible. This has been carried out with the utmost strictness. In an officer's cabin, for example, only the top on the writing desk and some loose items, such as windows with rattan seat on the sofa, book rack, etc. are made of wood.

Speed.
During the sea trial, 14.987 miles were achieved with 2086 HP and 167 revolutions per minute. The propellers are 3-bladed, have a diameter of 2500 mm, a pitch of 3750 mm, made of bronze and have fixed blades.

Boilers and machinery
Four water-tube boilers according to the Yarrow system supply steam required for propulsion, etc. Furthermore, two vertical auxiliary feed pumps and a Westinghouse soot-blowing pump are installed in the boiler rooms. The main machinery is designed for a capacity of 1000 IHP per machine at approx. 160 revolutions per minute. This number of revolutions has been increased somewhat hungrily.
The engine room also contains a vertical duplex steam pump, which can serve as a bilge pump, fire-fighting pump and scoop water pump and can also be used on the circulation space of the distilling equipment, the condenser of the ice machine and on the supply line of the evaporators; an electrically driven main centrifugal pump

Steering
These ships are the first warships to be equipped with a Flettner rudder. One of the most skilled English shipbuilders once called the steering gear the Achilles heel of a modern warship, and certainly not without reason, considering the many times on which this part has caused trouble in action. Because the Flettner rudder makes the entire steering gear superfluous, a weak point is eliminated. The Flettner mechanism is so simple that damage to it seems impossible, and it is so compact that a hit would also fix and damage the rudder stock. The mechanical connection from the steering position on the bridge to the steering gear is laid in the lower part of the ship, where it is completely protected. It is made with special care, mounted entirely mounted with ball bearings with numerous universal joints in between, in order to prevent the adverse influence of elastic changes in the shape of the ship in seagoing conditions.
Although a judgement about the installation can of course only be made after long-term use, the preliminary result is very favourable. The rudder can be placed from helm to helm in a few seconds, much faster than a steering machine could do. It reacts to every movement of the steering wheel, even if only 2 or 3 spokes of rudder are given, and can be moved "with one finger". It also showed its good functioning during manoeuvering and turning in the narrow harbour of Den Helder, even at extremely low speed. When the ship is stationary, the propeller water is sufficient to allow it to work: when turning while moored, the rudder can be placed 45° on board.

SOEMBA

Flotilla Leaders					
Name	Builder	Laid down	Launched	Commissioned	Fate
Flores	Mij. Voor Scheepsbouw en Werktuigbouw Fijenoord N.V. at Rotterdam	13 January 1925	15 August 1925	25 March 1926	12 November 1968 sold for scrap.
Soemba	- N.V. Wilton's Machinefabriek en Scheepsbouw at Rotterdam - Grangemouth & Greenock D.Y., Grangemouth 1945/1946 (reconstruction to aircraft direction ship) - Navy Yard Den Helder (Rijkswerf Willemsoord) 1950/1951, 1956, 1962-1964	24 December 1924	24 August 1925	12 April 1926	5 July 1985 sold for scrap

DEVELOPMENT

The two ships were designed by engineers L. Troost (1895-1966) and M.F. Gunning (1895-1972) and included in the 1924 budget. Flotilla-leader *Soemba* was laid down on 24 December 1924 at 'Wilton's Machinefabriek en scheepswerf' in Rotterdam and launched on 24 August 1925. The ship consisted of a steel construction with transverse and longitudinal frames. Over 81% of her length had a double bottom. Her armour consisted of 215.9 tons of nickel steel (Krupp).
It was a successful design with good seaworthiness due to her high forecastle. But also due to her good gun platforms and steering characteristics. Despite that, maximum speed was lower than the desired 15 knots. Also the ventilation, especially of the ammunition storage places was inadequate.

While serving in the Netherlands Indies, an aircraft could be embarked. Here the V-3, Fokker C-VIIw, reconnaissance plane that was delivered in 1929. (NIMH)

Technical data		
Displacement:	1,457 tons standard, 1,734 tons full load	
Length:	75,50 m (over all)	
Beam:	11.50 m	
Draught:	3.60 m	
Machinery:	2x Triple expansion steam engine, 4 Yarrow boilers, 2000 HP (Engineerd by builders)	
Speed:	2 shafts, 15 knots	
Range:	-	
Armour:	Deck 25-50mm, Command tower and ammunition chutes 25mm	
	Gunboat	Instruction ship
Complement:	136	143
Armament:	3x gun 15cm, 1x gun 7.5cm, 4x machine gun 12.7mm, 2x machine gun 7.7mm, smoke generator able to carry Fokker C.XIV-W floatplane	Initially disarmed later: 2x machine gun 40mm 2x machine gun 20mm.

Soemba was commissioned on 12 April 1926.

In 1935 the two flotilla vessels, *Soemba* and *Flores*, were classified as gunboats. In the years before the war its appearance hardly changed. The changes appear to be limited to equipment, such as whether or not to carry life rafts, paravanes and shooting targets.

Departure to the Dutch East Indies with Flores. *(NIMH)*

WEAPONS

The main armament consisted of three 15 cm Mark 7 guns; two on the forecastle and one on the campaign, mounted behind shields. These guns were manufactured by Bofors in the 1920s. In WWII, both ships were supplied with British munitions. The guns were extensively used during the invasions of Italy. By 1944, both ships had worn-out guns, which were replaced with ones removed from the cruiser *Sumatra*. Actual bore diameter was 14.91 cm (5.87") for all guns.

Technical data	
	Bofors 15 cm gun
Designation	15 cm/50 (5.9") Marks 6, 7 and 8
Date In Service	1925
Calibre	15 cm
Rate of fire	7-8 rounds per minute
Max Range with 103 lbs. (46.7 kg) AP	29 degrees: 23,200 yards (21,214 m)
Train rate	N/A
Elevation	N/A
Train	About +150 / -150 degrees
Total weight	20 tons

Eight Mark 6 guns were produced by Krupp, while twelve guns were made by Bofors. Mark 7 guns were all made by Bofors. Mark 8 guns were produced by HIH.

KANON v.15c.MN°7.

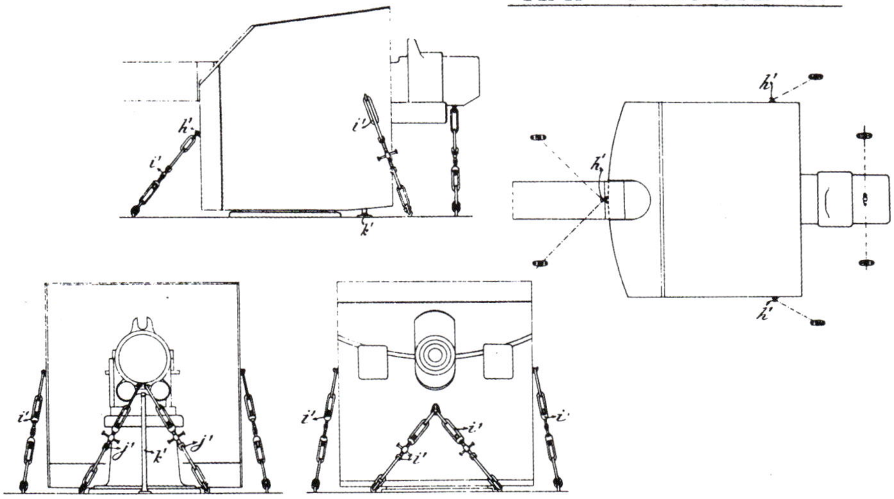

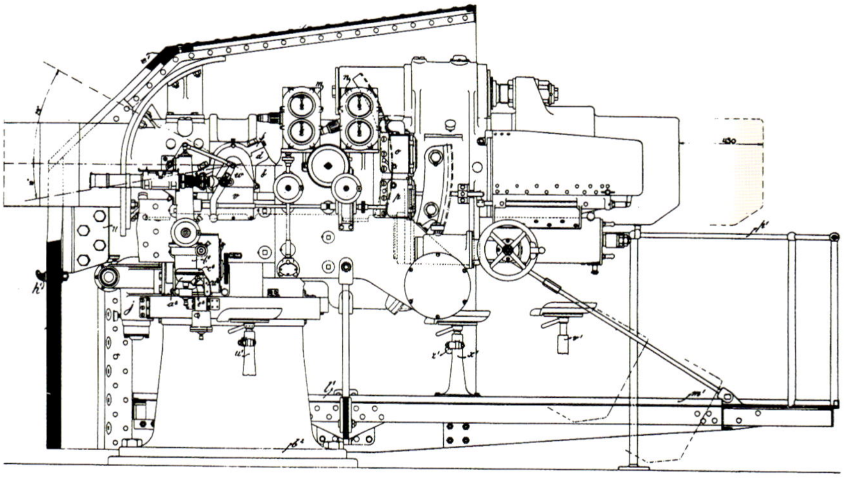

Note: Guns of Soemba *were without the platform.*

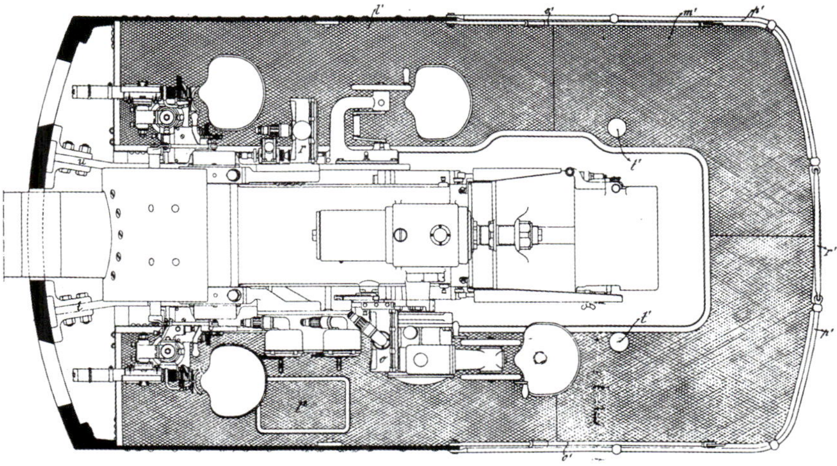

Secondary, one Bofors 7.5 cm gun for the AA role that was exported to the Netherlands and designated as the Mark 6. This gun was obsolescent by the start of World War II. Fired fixed ammunition.

Technical data	
	Bofors 7.5 cm gun
Designation	7.5 cm/60 (2.95") Mark 6
Date In Service	1928
Calibre	7.5 cm
Rate of fire	12 - 15 rounds / minute
Max Range with 14.3 lbs. (6,5 kg) HE	45 degrees: about 18,370 yards (16,800 m)
Train rate	Manually operated, only
Elevation	-5 / +95 degrees
Train	-360 / +360 degrees
Total weight	4900 kg

Also, four 12.7 mm Vickers machine guns. (The Vickers 12.7 mm were also known as the Vickers .50) It was a large-calibre British automatic weapon. Designed following the inter-war doctrine of using heavy machine guns as close-range anti-aircraft weapons on warships. The guns had 200 magazines per barrel with a 700-round per minute cyclic rate for each gun on the mount. Mounted 2 x 2 on bridge platform.

Vickers .50 machine gun

Further two Vickers 7.7 mm machine guns. These guns were operated by a three-man crew but typically required more men to move and operate it: one fired, one fed the ammunition, the others helped to carry the weapon, its ammunition, and spare parts.

The ship had also a fog installation.

Waterplane

The ship was able to handle a waterplane, being stowed on the deck aft. From time to time, she had a Fokker C.VII-W sea scout on the afterdeck. Sometimes, until 1928, also a plane of the type Van Berkel WA. In 1940, in the context of a trial, possibly a Fokker C.XI-W. Carrying a seaplane did, however, render the rear 15 cm gun unusable.

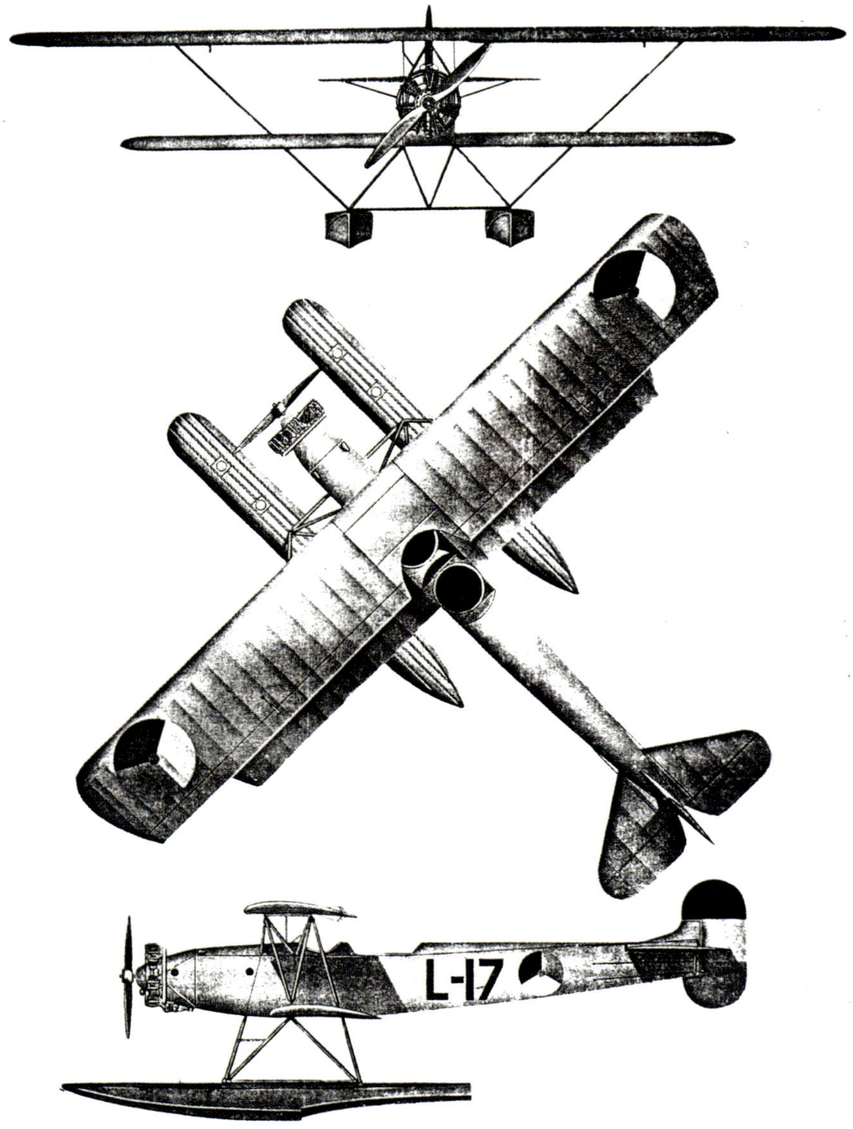

Primarily a Dutch Naval Air Service training aircraft, the Fokker C.VII-W was also used for reconnaissance duties.

An aircraft was only taken on board in the Dutch East Indies. The jettison of the C.VII-W was far from easy. It depended on the wind and sea conditions. Because the ship had to be stationary while the planes were being lowered, this meant a significant loss of time. Moreover, at that moment, they were an easy target for a submarine.

Only the Netherlands Naval Air Service flew the Fokker C.VII-W. The aircraft was never sold abroad. The first aircraft had an angular, not a rounded, underside of the vertical stabiliser. The pilot and observer sat in tandem, open cockpits. The wing structure was wooden with fabric and plywood covering, and the fuselage was of steel tube construction with fabric covering.

The first twelve of the thirty examples produced were sent to the Dutch East Indies, with the rest remaining in the Netherlands. The type was withdrawn from front-line service in 1940, but some machines remained active in the East Indies as trainers until the Japanese invasion in 1942.The C-VII-W was Fokker's first successful floatplane.

Technical data	
	Reconnaissance plane
Designation	Fokker C-VII-W
Date in service	1930
Length	9.98 m (32 ft 9 in)
Wingspan	12.9 m (42 ft 4 in)
Height	4 m (13 ft 1 in)
Empty weight	1,120 kg (2,470 lb)
Gross weight	1,620 kg (3,570 lb)
Engine	Armstrong Siddeley Lynx 7-cylinder air-cooled radial piston engine, 168 kW (225 hp)
Max. speed:	160 km/h (99 mph, 86 kn)
Crew	2
Weapons	Provision for a machine gun on a Scarff ring in the rear cockpit

Embarked on Soemba. *As early as 1924, the navy experimented with carrying aircraft on the aft ship.* (NIMH)

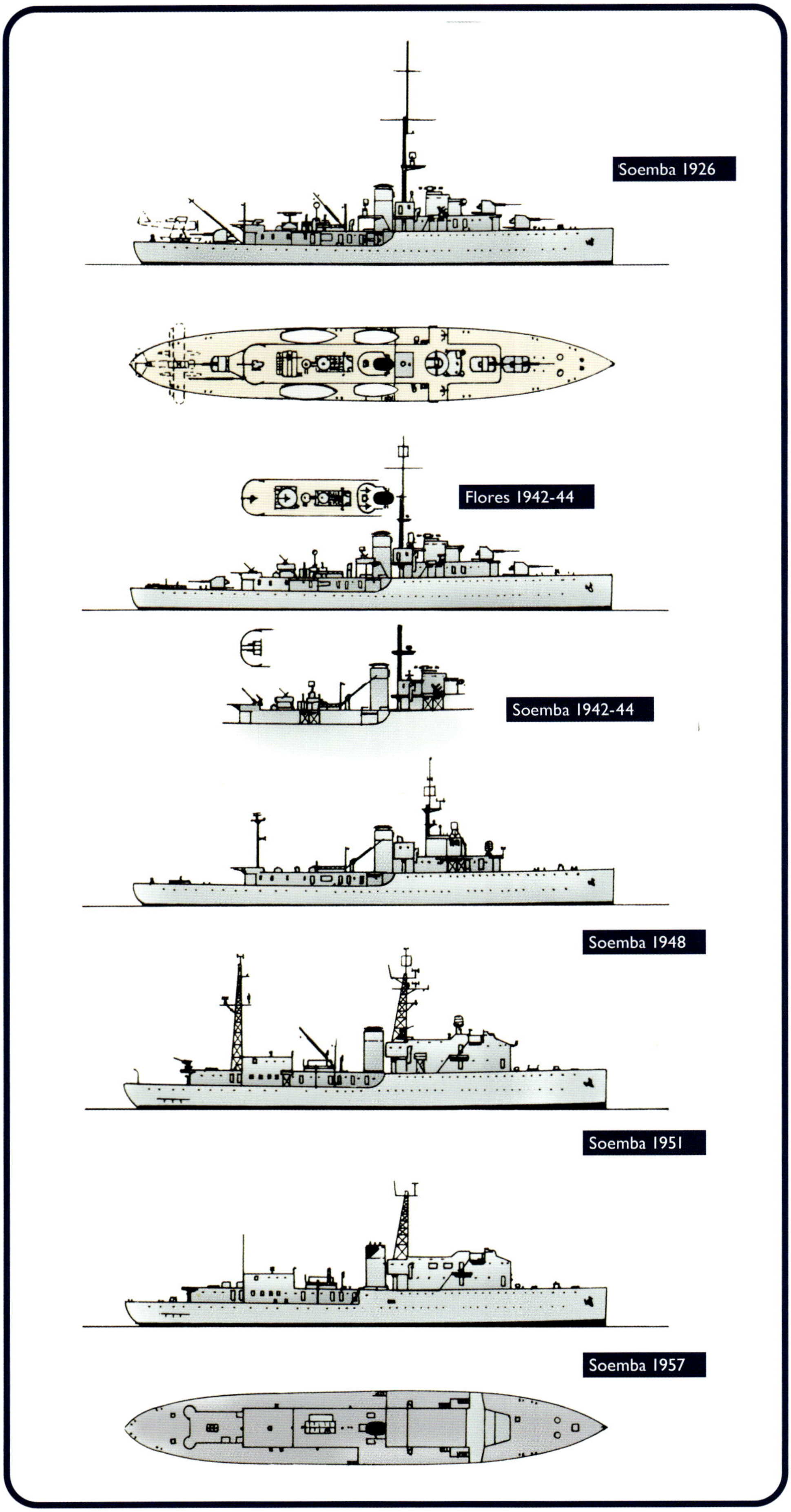

WARTIME MODIFICATIONS

End 1941/early 1942
Camouflage applied. This can best be described as a simple form of an 'angular disruptive pattern', which should be somewhat suitable for conditions with bright sunlight and moonlight (such as in the Dutch East Indies). On both sides the ship had the same pattern. *Soemba*'s scheme was different from *Flores*.

April 1942
Naval Dockyard in Bombay.
The mast was shortened and 8 life rafts were added. The aircraft loading boom, the searchlight on the mast and the range-finder aft were removed.

January-April 1943
Naval Dockyard in Bombay
The 7.5 cm gun is removed. Radar is installed and the ship is given camouflage colours for the Mediterranean Sea. Admiralty standard light tone (Superstructure and fore deck light grey, dark grey underneath)

Gunboats Soemba *(left) and* Flores *in WWII.*

May 1944
Naval Dockyard in Portsmouth
The radar is renewed and expanded, VHF radio communication equipment is installed. Camouflage applied: Admiralty disruptive

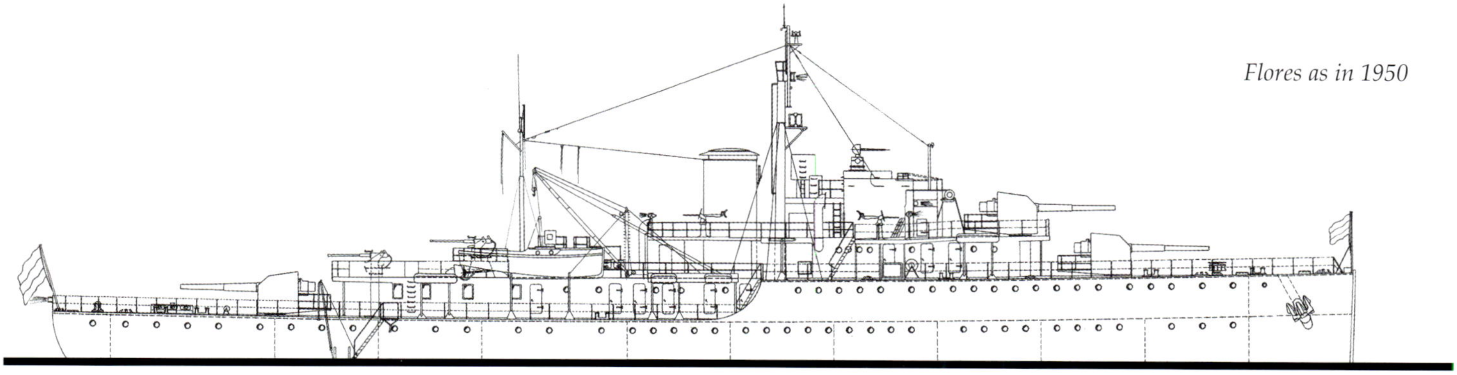

Flores as in 1950

June 1944
Plymouth
VHF radio communications equipment is improved

After 1945
The ship had a closed bridge. The bridge superstructure was also greatly enlarged.

Crew
Originally, the complement was 132/134 men (of which a maximum of 67 natives). In World War II this went up to 143 men. After her reconstruction to radar instruction ship in 1944 this became 185 men. After the war as aircraft direction ship in 1949 the ship had a complement of 164 men. As accommodation ship since 1964, she could house a maximum of 176.

It was not allowed to photograph ships in port during the war. But sometimes a rare colour photo emerges. Here, Flores *is showing her camouflage coat.* (NIMH)

Another camouflage scheme was applied to Soemba.

Armament In 1944:

- 3x guns 15 cm
- 1x Bofors 40 mm twin
- 4x Oerlikon 20 mm
- 4x 12.7 mm (2 on bridge wings and 2 aft)

All guns were manufactured by Bofors, Hollandsche Industrie-en Handels-maatschappij (HIH) and Krupp and used by the Netherlands Navy on cruisers and sloops (gunboats) built during the 1920s. They used horizontal sliding breech-blocks and had similar ballistics. Surviving ships were supplied with British munitions during World War II. The sloops *Flores* and *Soemba* were extensively used during the invasions of Italy and had their worn-out guns replaced with ones removed from the cruiser *Sumatra*. Actual bore diameter was 14.91 cm (5.87") for all guns.

On 26 August 1944, *Soemba* was laid up. The poor condition of the main battery in particular means that the ship can no longer be used for war purposes. Training courses were started on board, first to be constable and then training to be a sailor. At the end of October 1944, the Royal Netherlands Navy requested the Royal Navy to convert her into a radar training ship. This conversion took place from 21 May 1945 to 7 May 1946 in Grangemouth, near Edinburgh, Scotland.

Pre-war photo showing details of the midship.

The appearance of the ship has now been thoroughly changed. The armament is all gone, the mast has been moved forward against the still present armoured command centre, the original open bridge has been closed and the overhanging part of the fore deckhouse (where gun 2 partly stood) has been removed. Additions were a cabin. in front of the bridge and the entire arsenal of English radar that the Dutch navy had at that time, as far as it did not concern direct aiming equipment for weapons.

Main gun aft of Flores, *1944.* (NIMH)

RADAR

Visiting Rotterdam in 1948. Note the British radar and boat carried. (Collection T.L. Sander)

After the Second World War, the Royal Netherlands Navy had to be rebuilt.
In 1946, the first radar training courses started on board of *Soemba*. Sailors were trained for the Radio Distance Peiler Plot (RAPP) certificate, followed by cadre training courses. From 1949, these cadre members were given their own service department: the Navigation Combat Information Service (NGID).

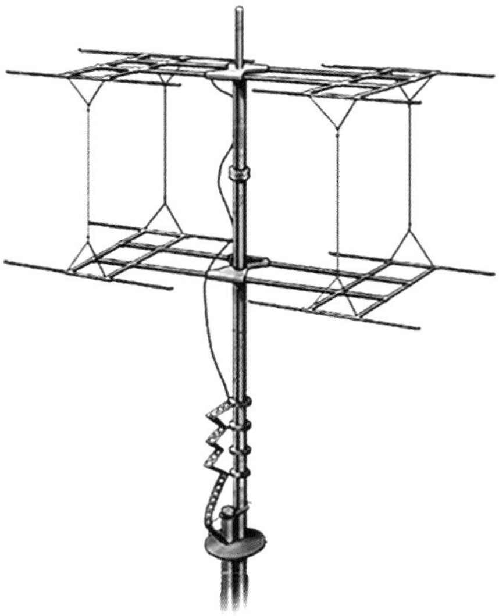

Type 281B

Several radar systems were installed, including:

- Type 281B long-range airborne radar system on the forward mast
- Type 243 IFF interrogator for aircraft
- Type 253 IFF interrogator in the top of the mast
- Type 271 surface scanning radar above the bridge
- Type 291 air surveillance radar on the forward mast
- Type 277 radar for height determination
- Type 268 submarine detection radar on a mast amidships
- Type 293 radar on the main mast

With this configuration the ship was fully adapted for her new role as an aircraft control ship, where she could escort fighter aircraft during exercises and missions at sea.

Type 281
On the mast between the bridge and the funnel is the large, rectangular antenna of the Type 281 long-range air warning radar. The main air warning radar of the Royal Navy for large ships in the war.

The Type 281 is a direct descendant of the Type 79, the first radar that the Royal Navy developed and put into service in the second half of the 1930s. The single-mast version was called Type 281B and had a range of 120 nm (about 222 km)
The *Soemba* carried the last version of this type, the 281BQ, or BQ81 for short. In the Netherlands navy, the Type 281 was only carried by the two aircraft carriers that were taken over from the Royal Navy after the war. However, they had antennas from an earlier version, Type 281B, where the signal supply and discharge still took place via coaxial cables and it was not possible for the antenna to rotate continuously. In the Type 281BQ, the cables were replaced by a coaxial coupling that made this possible.

Type 243
Directly above the Type 281B was a Type 243 (IFF) interrogator for aircraft. This antenna rotated with the Type 281 and transmitted a signal that was returned by the (friendly) aircraft and could be included on the display of the Type 281.

The Royal Netherlands Navy and Development of Radar

Neither a single nation nor a single person can say that the discovery and development of radar technology was his (or its) own invention. One must see the knowledge about "Radar" than an accumulation of many developments and improvements, in which any scientists from several nations took part in parallel.
At the outbreak of WWII, different radar equipment was developed in the USA, Russia, Germany, France and Japan.

The research regarding radar and its development in the Netherlands was done by jhr.ir. J.L.W.C. Von Weiler and his team. It was interrupted by the outbreak of World War II in May 1940. At that moment four prototypes of this M39 electrical listening device were made. The Ministry decided to start small-scale production at NSF in Hilversum. Ir. Max Staal, who was performing his military service at the Laboratory for Physical Warfare, was appointed liaison officer to maintain contacts with industry.

Of the four prototypes, one was installed in The Hague and during the Invasion of 1940. used to locate German aircraft. A device on the roof of the Waalsdorp laboratory was dismantled and destroyed. The remaining two dismantled units were shipped to England at the behest of Ir. Piket.

Ir. Von Weiler and Ir. M. Staal managed to escape on the destroyer HMS *Wessex* to the United Kingdom, as the Germans invaded the Netherlands, taking confidential information with them such as designs and drawings. Shortly after, they were placed in *HM Signal School* at Portsmouth where equipment was built. The development team was astonished to learn that the Dutch used only one antenna for transmitting as well as receiving, as British radars so far used separate antennas.
During WWII, all activities concerning radar were halted in the Netherlands. Von Weiler and Staal worked together with their colleagues in the United Kingdom and took advantage of this close cooperation.

The Dutch systems were adapted to serve as rangefinders for the two Hazemeijer fire-control systems of two twin 40 mm Bofors machine guns, which the Royal Netherlands Navy was using on its new ships. Double Yagi antennas of British design were used as antennas. The "Dutch" radar systems – Radio Direction Finder Type 289 (RDF 289) – were installed on the destroyer HNLMS Isaac Sweers.

After the war
After the move of the Naval Staff from London to Den Haag (The Hague) on 1 September 1945, the most important task was to get the navy 'back on its feet' again along with reviving the national economy, by involving Dutch industry as much as possible.
The country had suffered greatly under the occupation. The Royal Netherlands Navy was reduced in quantity and in quality as many ships had been destroyed. Those ships fortunate enough to return to the national harbours were often obsolete.

The initial plans were over-ambitious and had to be reduced during the course of the planning process. The original plan involved 3 Task Forces, but was then honed down to 1 Task Force consisting of 1 carrier, 2 cruisers, 12 destroyers and 8 submarines being in service in 1953. Rebuilding the Navy was obviously not an easy task. There was a lack of money and basic materials, and many shipyards were damaged.

Prof.jhr.ir. J.L.W.C. Von Weiler: Pioneer of Dutch radar development

Continuation was secured by buying surplus ships from the Allies. At the same time, new plans were developed for the construction of new ships by Dutch shipbuilders and their subcontractors.

In 1946, the department of 'Test and Development' of the 'Marine Radiodienst' in Oestgeest started with the development of an integrated plan for surveillance and fire-control radar systems on board the ships of the new fleet, under the management of Von Weiler. From the industry sector 'Philips Telecommunicatie Industrie' (PTI) in Hilversum and Signaal became involved in the development of this integrated plan for radar systems.

The PTI in Hilversum

Type 271

A Canadian developed 1944 surface search and navigational radar for the Royal Navy. It was a compact 'cheese' radar, often fitted in coastal forces craft. One antenna was for receiving and the other for transmitting. The antennas were steered using an automobile steering wheel from an operator's station directly below. This arrangement minimised the length of the coaxial cables used in this very early microwave set.

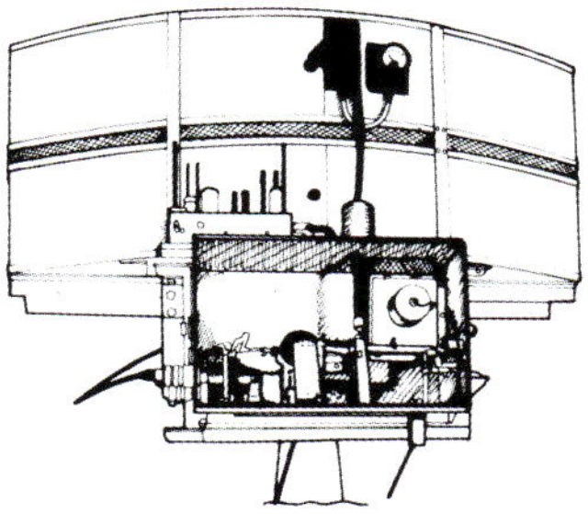

Type 271

Type 291

This radar was an British early warning radar and became operational end of 1941. It was widely used on small ships throughout WWII. The antenna was still turned manually with a crank.

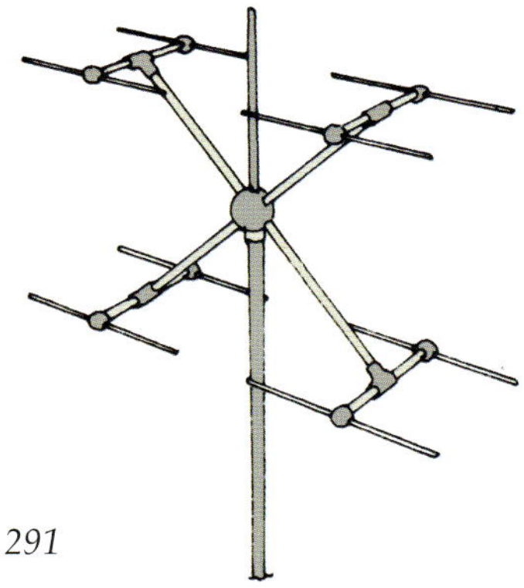

Type 291

Type 277

On the deck in front of the bridge is a Type 277 antenna, considered by many to be the most successful English radar of the war. Originally intended as a sea warning radar for smaller ships, the device with an elevating dish also appears to be suitable for altitude measurement of aircraft and detection of low-flying aircraft. With the latter it is an addition to the Type 281, where that is a weak point. The result is that both are often found in combination on large ships. The last version of the radar that was developed during the war was on *Soemba.*

Soemba *in 1949.*

The Type 277 may have been an excellent radar, but it was still difficult to get the pencil-like beam on the target. To overcome this problem, the same installation was equipped with a stabilised 'cheese'-shaped antenna that covered a larger vertical area. This became the Type 293 and was placed on top of the vertical rack on the stern at *Soemba.*

Type 268

A search, navigational and target-indication radar developed in Canada for the British Admiralty in 1944.It was used for detection of the snorkel of a submarine.

Type 293

A surveillance and target designation radar. The radar is equipped with a so-called "cheese antenna" forming a fan-beam antenna pattern. It acquired that name because it looked like a block of cheese cut in half. The Type 293 radar system first saw service in 1944

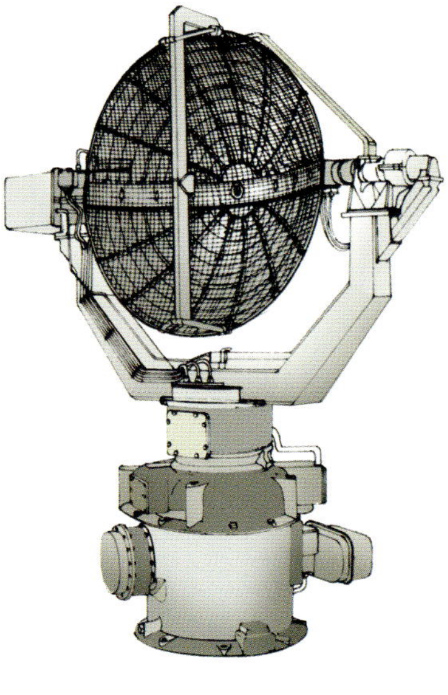

Type 277

Below: *Visiting Rotterdam*

On 23 June 1949, Soemba participated in a Road Safety campaign featuring a symbolic youth traffic trainer named 'Bob'. Here arriving in Amsterdam were 'Bob' disembarks. (National Archives)

Youth traffic trainer 'Bob' leaving the ship. On the quay, spectators with banners gather. *(National Archives)*

Soemba with large pennant number in 1947. (National Archives)

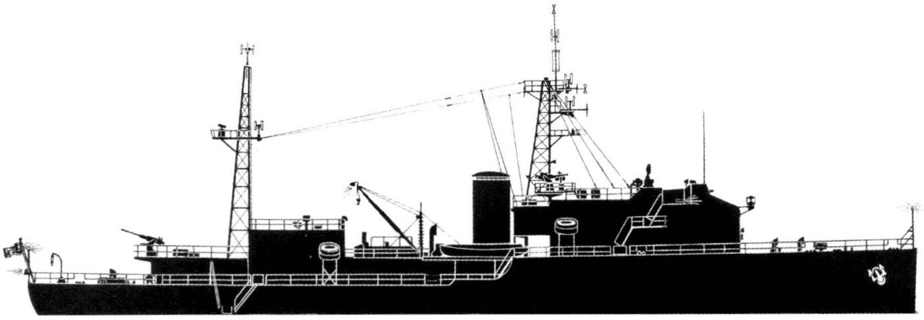

In 4 October 1949, *Soemba* was decommissioned as a radar training ship, because sufficient training facilities were now available on shore. However, it was decided not to retire the ship, but to convert her into an aircraft control ship. This meant a new role in guiding and coordinating air operations. An important role within the modernizing post-war navy. The conversion took place at the navy yard in Den Helder. The armoured conning tower being removed and both the forward and aft superstructures being enlarged. The original masts were replaced by lattice masts, which were better suited for carrying radar systems. In addition, the ship was given a new armament, consisting of:

- Two 40mm Bofors guns aft
- Two 20mm Oerlikon guns, mounted on either side of the bridge

She was recommissioned on 1 June 1951. And whether it is due to the greatly enlarged superstructure is not clear, but above wind force 4 BF the ship normally no longer leaves the harbour.

The radar configuration has been modified for the new function, which means that at least the antennas that are not in any way related to aircraft, such as the Type 271/272 and 268, have been removed. In addition, the Type 291 air warning radar has also been removed. The rest is still there, but sometimes in a (slightly) different place. For example, the Type 277 has ended up on top of the bridge and the Type 242 interrogator that was previously located there has been moved to the upper part of the foremast. Some more things have been added: a medium frequency direction finder (MF/DF), a talk-between-ships (TBS) antenna and behind that a Type 253 transponder.
The conversion to an aircraft direction vessel did not mean that the instructions on board were over. The radar-related training had been moved to shore, but the final part of the training usually involved a placement on board *Soemba*.

Soemba's role as an aircraft control ship ended on 30 November 1954 when she was decommissioned and laid up. In 1956 all armament was removed

(NIMH)

After reconstruction berthed in IJmuiden 1951 (National Archives)

HNLMS Soemba

12 April 1926 - 7 June 1985

Commissioned	First time Suez Canal	Last foreign port call	Decommissioned
12 April 1926	1926	Portsmouth 1951	7 June 1985
Radar instruction vessel	**Aircraft direction vessel**	**Accommodation ship**	**Fate**
1946	1950	1956	Scrapped

Commanding Officers		
From	To	
1926	1927	J.F. Osten
1927	1929	J.J.v.d.Linden
1931	1932	C.J.v. Asbeck
1932	1932	A.C.W. Kerlin
1933	1933	H. Vreedenbergh
1933	1933	L.A.C.M. Doorman
1933	1934	W. van Lier
1934	1934	R.P. van de Wetering
1934	1936	W.H.C. Hoog
1936	1937	C.H. Brouwer
1937	1937	J.C.A. Scholte
1938	1939	J.J. Wichers
1939	1939	M.M. Merens (temp.)
1939	1940	G.A. Berg
1940	1941	J.C. Cornelis
1941	1943	P.J.G. Huijer
1943	1943	J.J.M. Sterkenburg
1943	1943	A. van Miert (temp.)
1943	1943	P.A. de Boer
1943	1944	H.H.L. Pröpper
1944	1944	J.G. Broekhuysen
1944	1944	P.P. Schotel
1944	1944	T.H. Brunsting
1944	1945	S.L. de Jong
1945	1945	H.A. van Oorde
1945	1946	J.J. Hoogendoorn
1946	1947	C.H. Frijlink
1947	1948	J.B. Kennema
1948	1949	J.G. Stegeman
1949	1949	M.J. Vos
1951	1952	N.J. Jongewaard
1952	1953	P.W.C. de Vos
1953	1953	S. van Sorge (temp.)
1953	1954	J.D. van der Star
1954	1954	W.F. van de Valk Bouman
1958	1958	D. Teer
1958	1960	A.E. de Jager
1960	1960	C. Huijsman
1960	1960	J.M. van Eck Poppe (temp.)
1960	1960	C. Huijsman
1960	1961	W. Enterman
1961	1961	J.M. van Eck Poppe (temp.)
1961	1963	H.P. van der Lelie
1964	1964	F.G.A. Woudstra
1964	1966	A. Tortike
1966	1968	F.Th. de Bree
1968	1970	W.W. Bak
1970	1972	W.K. Verboeket
1972	1974	H.P. van der Lelie
1974	1978	W.J. de Bruyn
1978	1979	E.C.A. Tekke
1979	1983	J.O. Geertsma
1983	1985	J.G. Vos

On 10 August 1951 a shipscrest was allocated to radar instruction vessel Soemba. It portrays the insignia on the radar operators certificate. The Latin motto: Sic Iaco Sagittas translates 'Thus I throw arrows'.

When she became an accommodation ship for diver training she received in 1968 a new crest created by G.A. Bontekoe. It shows the herd of Sandalwood horses that roam the island of Soemba, Both grey and dark types occur in this breed.

OPERATIONAL HISTORY

Building yard number 304, *Soemba*, at Wilton in Rotterdam. (Collection T.L. Sander)

In 1924 it was decided to build two flotilla leaders for service in the Netherlands Indies (Indonesia). On 24 December, she was laid down at Wilton's Dok- en Werf Maatschappij.

Wilton's Dok- en Werf Maatschappij
An independent company from 1854 till 1929. At first it was simply known as 'Wilton'. In 1921 the name became: 'Wilton's Dok- en Werf Maatschappij NV', the equivalent of 'Wilton Engineering and Slipway Company'. Wilton started as a traditional smithy and expanded in machinery. It specialized in ship repairs, but also became a shipyard. At first only for fresh water ships. In spite of multiple name changes the company was commonly referred to as 'Wilton'. In 1929 a merger with Fijenoord led to a new company known as Wilton-Fijenoord. Presently, the shipyard of Wilton-Feijnoord is part of Damen Ship Repair Rotterdam.

On 24 August 1925 she was launched and commissioned on 12 April 1926, followed by preparations for the journey to the Netherlands Indies together with *Flores*. An accident in the engine room seriously injures the stoker-oiler to the head. On 15 June, after a visit of the vice admiral both flotilla leaders depart. They formed a keel line at the roadstead and then steamed past the squadron, which consisted of the armoured vessels *Maarten Harpertsz. Tromp* and *Jacob van Heemskerck*, the submarines *O 9* and *O 11*, and the torpedo boats *Z 7* and *Z 8*. Escorted by the latter two ships and several aircraft, the *Flores* and *Soemba* put to sea at 12:30 via the Schulpengat. The ships exchanged the customary tributes with the minelayer *Douwe Aukes* off Scheveningen and set course for Dover Strait via the lightships *Maas, Noordhinder*, and *Sandettie*. The next day the ships pass Dungeness at 10:30 a.m. and follow the English coast. At 7:00 p.m., they pass the lightship *Owers* and at 11:00 p.m., St. Catherine Point (Isle of Wight).
After sailing 1.5 hours in a thick fog on 17 June, they reach Cape Finisterre on 19 June. At a great distance, the British cruiser *Concord* is being scouted.

Visiting Seville (22 - 26 Jun). Problems arise with the ammonia ice machine on *Soemba*. A serious leak causes ammonia to escape. Because everything is happening so quickly, the crew on board are unable to close the watertight door, and the ammonia spreads into the main engine room. The stoker-oiler on duty has to be rescued by crew members who enter the engine room wearing gas masks.

The ships enter the Strait of Gibraltar and pass the Almina Point lighthouse on 27 June 6:00 a.m. The following day, they conduct joint exercises with the paravanes, for which they stop temporarily and sail at various speeds.

Visiting Tunis (1 - 5 Jul.) before heading to the Suez Canal. A replenishnent stop was made in Suez (12 - 14 Jul.). On 16 July they pass Daedalus Reef and then the Tropic of Cancer. On 19 July the ships reach the lighthouse of Quoin Island, east of Jebel Zukur. During the voyage through the Red Sea, 10 men aboard the *Flores* have to be relieved because they are suffering from heat exhaustion.

Lightship Noord-Hinder *located about 90 kilometres off the coast of Hoek van Holland. Moored at the Noord Hinder sandbank and therefore named* Noord-Hinder. (NIMH)

Daytime temperatures in the engine room reach 56 degrees. The ships sail through the narrow Babal-Mandeb Strait and steam out of the Red Sea. Visiting Aden (20 - 22 Jul.) for replenishment. On 23 July the wind picks up. The number of strokes is reduced to 100. The next day on the west coast, the wind increases to 7-8. The ships roll heavily, but they hold their ground. All relevant openings are closed and covered. In the aft ammunition storage area of *Soemba*, eight torpedo warheads become detached, each of which is secured again. (These warheads are being transported as cargo). The ships approach the island of Socotra and constantly switch between fore and aft boiler rooms, because the temperatures in the boiler rooms are rising very high.
Soemba experiences a malfunctioning pump, causing a pipe to become blocked and a flange to burst. The engine room crew manages to repair the pump problem and installs a new flange. Two days later the weather improves, allowing some covered skylights and windows to be opened to freshen the ships' interiors.

On 31 July, the ships pass the British cruiser *Delhi*. The next day they arrive in Colombo where they moor directly behind the British cruiser *Effingham*. Three days later a Japanese squadron arrives in Colombo. Vice Admiral Yamamoto commands this squadron. The commanders of both Dutch ships would not have suspected, when they received the admiral on their ship, that they were dealing with the later architect of the attack on Pearl Harbor.

Below: *The 'Training Squadron' is moored in Makassar harbour opposite the Royal Packet Shipping Company office in 1928. The ships are, from left to right: the destroyers HNLMS* Panter *(front),* Lynx *(front),* Flores, Soemba *(back),* De Zeven Provincien *(armoured ship), and finally* Java *(cruiser). (NIMH)*

Harbour of Tanjong Priok (Batavia), where the squadron was moored. In the foreground is HNLMS Sumatra, *behind it* Soemba, *and beyond that, several submarines. In the background, the volcanoes Gedeh (left) and Salak (right).*

Departing from Colombo on 5 August. The prescribed honours are exchanged with British and Japanese warships. The flagships *Effingham* and *Izumo* hoisted signal flags wishing them a 'bon voyage'. On 10 August, the lights of Poeloe Bras and Poeloe Klah were spotted. Soon after, they arrive at Sabang (the port of Poeloe Weh in North Sumatra).

Now they are under command of CZMNI (Commander-in-Chief Naval Forces Netherlands Indies). It is also the location for replenishment.
Visiting Tanjong Priok (18 Aug - 13 Sep) before heading to Surabaya where they were assigned to 'Oefeneskader Kleine Soenda-eilanden' (training squadron) The Task Group also includes the cruiser *Java*, several destroyers, and submarines. On 20 October, they arrive in Makassar were the Task Group is disbanded.

After short maintenance in Surabaya, *Soemba* takes over patrol duties in the Sunda Strait from the minelayer *Krakatau* on 2 December. After her release on 11 December, she returns to Surabaya for repairs.

After some exercises near Palembang in February, *Soemba* is reunited with *Flores*. Visiting Samboe (7 - 8 Mar), Singapore (8 - 14 Mar) and Saigon (17 - 23 Mar.) before arriving in Tanjong Priok on 28 March. There she joins a training squadron for a journey around Sumatra (1 - 29 Apr).

On 3 May *Soemba* joins the Task Group for exercises in the western Java Sea. After which she escorts government steamship *Wega* (with the Governor-General on board) to the Moluccas. On 10 October a rendezvous with the Task Group was made to join. From 27 October to 1 January the ship receives maintenance, after which the ship is kept "ready for departure within 24 hours" for the possible protection of Dutch interests in China. Therefore, *Soemba* only exercises in the immediate vicinity of Surabaya. On 9 February the *Soemba*, minelayer *Hercules*, submarine *K X*, and several aircraft search in vain for missing crew members of the government steamship *Zeemeeuw*, which capsized and sank that day.
In August, the readiness period is extended from 24 hours to eight days. *Soemba* executed several exercises in September and October. In the north of the Kangean Archipelago, she serves as an aircraft mothership. On 26 October she returns to Surabaya for exercises and maintenance, for the rest of the year.

In January 1929 she is in Tanjong Priok to attend the visit of the Britisch cruiser *Kent* and the Australian submarines *Otway* and *Oxley*.
On 13 March, she arrives at Sabang to witness the arrival of Dornier aircraft *D-21* and *D-23* flown in from the Netherlands and the visit of the German cruiser *Emden*. The aircraft arrived on 15 May, and two days later were escorted out with gun salutes by *Soemba*. This was followed by exercises and recruitment visits to locations on the north coast of Aceh. Returning to Surabaya at the end of the month for an overhaul while *Flores* takes over her duties.

1930s
It would take until 1 April 1931 before *Soemba* is commissioned again. Now she is attached to the artillery instruction-ship *Pelikaan*. She departs on 14 May from Surabaya to Kebatu to assist the cruiser *Sumatra*, which had struck a reef near Kebatu. During the rescue operation,

the *Soemba* lost an anchor and chain, which were later recovered by minelayer *Krakatau.* Returning to Surabaya on 20 May.
On patrol in October, she brings up several foreign fishermen who have committed beach robbery, to Kota Baroe. By the end of the year she returns to Surabaya for maintenance. The bad economics and moreover cuts on wages is resulting in social unrest among the sailors and employees of the shipyard. It leads to tensions and *Soemba* is sent on patrol Tarakan in February.
In June and July, she executes exercises near Surabaya before departing for a 'showing the flag' journey to the east part of the archipelago, visiting Ambon, Boela, Manokwari, Geelvinck Bay, the Sermata Islands, Mapia, and Banda Neira. Returning in September. In November 1933 and June 1934, *Soemba* makes voyages to the Karimata Strait, the west coast of Borneo, and the islands in the southern part of the South China Sea to display the flag and monitor foreign fisheries.

In August 1934, she goes to Tanjong Priok for exercises. Here, she is assigned to escort government steamship *Rigel* (with the Governor-General on board) on a voyage to the Moluccas. Returning on 20 October. In Decenber, *Soemba* supervises foreign fishing in the Gulf of Tomini (Celebes). During this time, the ship exercises and conducts preparatory, night-time and combat gunnery exercises in the Madura Strait. A training voyage to Tanjong Priok is also made.

In 1935,* Soemba *is classified as a gunboat.

In September *Soemba* patrols in the eastern part of the archipelago. After short maintenance in Surabaya she conducts several gunnery exercises in the area. By August 1937 *Soemba* and the *Flores* take turns providing guard duty in this part of the archipelago until she is relieved for maintenance in September. In May 1938, she takes up guard duty again in Riouw archipelago. In September takes part in a fleet review at Surabaya.

In August 1939, *Soemba* is assigned to guarding the Sunda Strait, as are the destroyer *Banckert*, the submarine *K XV*, two ships of the Governmental Navy (the *Valk* and the *Gemma*), and a group of large aircraft. She departs Tandjong Priok on 5 September to monitor shipping in compliance with the neutrality proclamation issued on 3 September. After 3 weeks she is relieved by *Flores* and goes to Surabaya were she is assigned to protect minelayers. On 12 December she returns to guard Sunds Strait. During this period, the ship is also used to train gunnery spotters.

1940's
The *Soemba's* onboard aircraft is damaged and remains at the Priok Naval Air Station and will not be replaced. After arrival in Tandjung Priok, the damaged aircraft is loaded onto the ferry for transport to Surabaya.

On 18 March *Soemba* takes over the Riouw Archipelago surveillance from the destroyer *Witte de With* and conducts patrols to and around the Anambas Islands, the Natoena and Tambelan Islands, and on the west coast of Borneo between Tanjong Midai, Great Natoena, Pontianak, the Lingga Group,

Naval port of Surabaya. In the foreground is Flores. *(National Archives)*

On patrol in the interbellum. (NIMH)

and Karimoen. The ship carries a Fokker V aircraft. After a few weeks she is relieved by the destroyer *Van Nes*.

During the German invasion of the Netherlands, on 10 May 1940, *Soemba* is stationed at Poelau Samboe and is charged with guarding the Riouw Archipelago. On 12 June she is relieved by the destroyer *Van Nes* and sails to Surabaya for maintenance. Departure from Surabaya on 5 August. *Soemba* heads for the waters to the waters around Madura for training of remote observers and for combat exercises.

On 20 September, she sails to the vicinity of Bima (Sumbawa) for a exercise along the northern coast of the Lesser Sunda Islands. Returning to Surabaya on 26 September to replenish supplies before proceeding to the Riouw Archipelago and taking over guard duty from the minelayer *Krakatau* that needs repairs (as a result of running aground). In November, she is ordered to escort two large tin dredgers, towed by NISHM tugs, from Manggar (Biliton) to Pangkal Pinang (Banka). The minelayer *Gouden Leeuw* temporarily takes over guard duties. After her return to Surabaya, in December, *Soemba* acts as a training ship for sailors, stokers, and quartermasters, and for training gunners and gunnery spotters.

The Fokker waterplane stowed aft on the deck.. (NIMH)

On 24 January, *Soemba* is is active in the Westervaarwater, at the welcome of the minelayer *Willem van der Zaan*, which arrived from England via Cape Town. From 24 February to 24 March she is assigned to Riouw Archipelago surveillance, after which she again acts as an instruction ship.

In 1941, an ASDIC system was installed; the gunboat was the first ship in the Netherlands Indies with such a system.

When war with Japan breaks out, on 8 December 1941, the ship is assigned to guard the Westervaarwater, replacing the *Soerabaja*, which was operating against Portuguese Timor. On 15 December she guards *Gouden Leeuw* and *Krakatau* during minelaying operations in the eastern entrance to the Maduro Strait. After wich she returns to guard Westervaarwater until she is relieved on 24 December. The next day she makes a short stop in Surabaya and takes up patrol duties in Riouw Archipelago. A leak develops in the high-pressure pipeline at the southern entrance to the Banka Strait, and the vessel goes to Tanjong Priok for repairs.

Arrival at Riouw Strait on 31 December. The commander is charged with organising air defence, which proves to be very useful for Singapore. The British defenders of Singapore are informed in a timely manner of the Japanese attack waves.

On 9 January 1942, *Soemba* destroys 8 floating mines. On 13 January, Japanese bombers attack the ship, but the *Soemba* is not hit. The next day the ship receives orders to return to Tanjong Priok.

On 19 January, she departs of Tanjong Priok for Lampong Bay to escort the British troop convoy MS 2, entering the bay, which consists only of the *Aquitania*. This 45,000-ton ship departed Sydney on 10 January, under escort of the Australian heavy cruiser *Canberra*, with 3,456 passengers on board. In the Indian Ocean, the convoy protection is reinforced by the cruiser *Java* and the destroyers *Evertsen* and *Van Nes*. The next day, the ships arrive at Ratai Bay, on the west side of the Sunda Strait. Here, the troops disembark and board seven smaller ships (six KPM and one British) to be taken to Singapore. During the transfer of the troops, the *Soemba* and the *Van Nes* are part of the naval force providing cover for the operation.

Oosthaven 20 February 1942.

On 21 January, *Soemba* is deployed to guard Tanjong Priok and to escort convoys in the Sunda Strait. One of these is the KPM motor vessel *Straat Madoera*, which has run aground on the Catarina Rocks. By 28 January, the merchant ship is freed from the rocks and *Soemba* departs for Lampong Bay to escort a convoy expected there. During the evacuation of Oosthaven (South Sumatra), on 1 February, *Soemba* escorts merchant ships via the Sunda Strait to the Indian Ocean.

The gunboat is near Oosthaven on 16 February. A KNIL (Royal Netherlands East Indies Army) force and Australian soldiers who have escaped from Singapore are in Oosthaven. The latter are completely demoralised. The Commander of Maritime Resources, the commander of Oosthaven, and the harbour-master report on board and report the chaos caused by the Australian troops in Oosthaven. The Australians started looting, forced their way into the KPM office, forced entry into the safe, destroyed everything they could find, and started drinking. Dutch naval officers were also threatened. Order is only restored among the Australian troops when Lieutenant Huijer stops bringing in evacuation ships and threatens to open fire on the harbour. During the following days, *Soemba* provides assistance in the evacuation of the troops to Java and Lieutenant Huijer takes part in the destruction of five ammunition depots, buildings and sheds, four trains with fuel supplies, dozens of torpedoes, and loaded pirogues. The British destroyer *Electra* and the Australian corvette *Burnie* (later HNLMS *Ceram*) also contribute by shooting several old hulks into the ground. With assistance from *Soemba*, the *Burnie* is able to remove five Spitfire engines and equipment from the Royal Air Force. Lieutenant Huijer, who is on shore, nearly runs into a Japanese column. The Japanese occupy Oosthaven. Lieutenant Huijer tries to contact the resident, who lives in Tanjong Karang, but it remains an attempt.

On 22 February, Lieutenant Huijer dispatches a Norwegian and three British tankers to Fremantle. By the time Japanese aircraft appear on the scene, the tankers have already disappeared.
Before leaving, on 24 February, she fires 10 salvoes from its 15cm guns at the storage areas and emplacements of the now-occupied Oosthaven. The ship leaves the bay and steams toward the Java Wall.

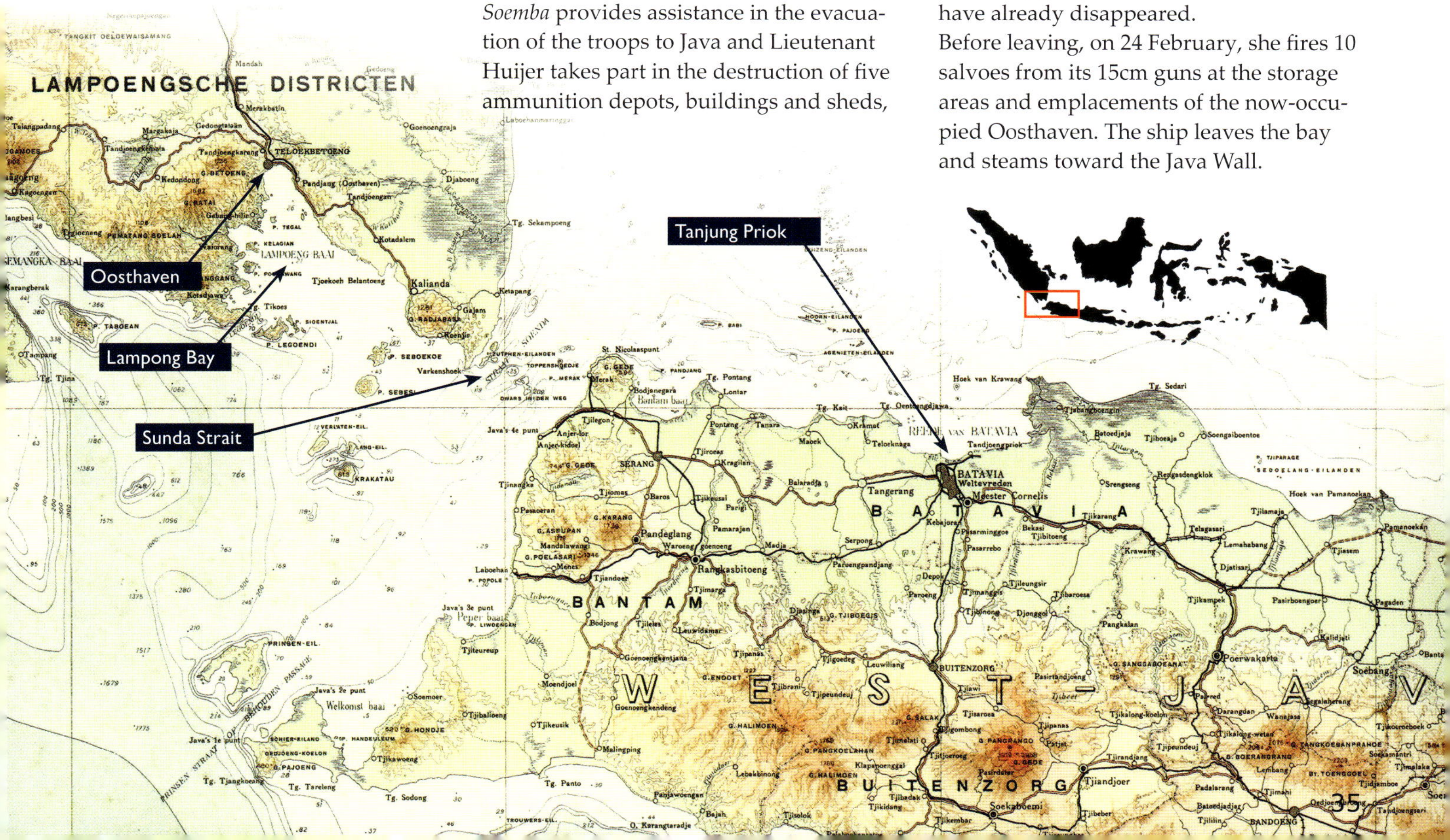

Soemba and *Sirius*, *Fazant*, and six Australian corvettes are moored at Merak on 27 February when three Japanese bombers attack the *Soemba*, during which one of the bombers is shot down. During the bombing, fifteen bombs fall in the immediate vicinity of the ship: five one metre starboard of the forward ammunition depot, five one metre port of the aft ammunition depot, and five nine metres aft of the ship. Nevertheless, *Soemba* sustains only minor shrapnel damage to the superstructure and hull. Unfortunately, one fatality is reported.

Near Merak, on 27 February, orders are received to depart immediately for Veeckens Bay on the west coast of Sumatra, where the tanker *TAN 8* (*Petronella*) is stationed. But *Soemba* has only 30 tons of fuel oil left on board. She arrives on 1 March and loads oil from *TAN 8*. Orders are received to protect the tanker. Departure from Veeckens Bay on 7 March for a rendezvous with the merchant *Siberg* (KPM) to proceed to Emmahaven together and then steam on to Colombo. The next day she reaches the rendezvous, but the *Siberg* is not encountered. *Soemba* moves back and forth for some time until she receives a message that the fight against Japan must be stopped. The submarines, gunboats, and the *TAN 8* are to proceed to Colombo. En route, the *Soemba* attempts to contact the *Siberg*, but due to a lack of fuel, it is abandoned and course is set to Colombo. (The *Siberg* remained in Wijnkoopsbaai and fell undamaged into Japanese hands.) Because of technical problems the speed drops to 10 knots. Arriving in Colombo on 23 March, and placed under command of Commander-in-Chief East Indies (Eastern Fleet) Admiral

Despite orders that Siberg *should have departed on 6 March, taking evacuees if possible, the ship remained in port. In a panic, the entire crew, including the captain, abandoned the ship. No measures had been taken beforehand to scuttle the ship in the deep bay. Consequently, the ship fell into Japanese hands undamaged. It was put into service as an ammunition transport ship, named* Sumerusan Maru. *When loading in the port of Surabaya, on 16 March 1943, ammunition exploded, causing irreparable damage to the ship.* (MarHis Data)

Tanker Petronella *served as* TAN 8 *with the Royal Netherlands Navy between 1939 and March 1942 in the Dutch East Indies. She was released by the RNN in April 1942 and leased to the Royal Navy. On October 15, 1944, she struck a mine in the Gulf of Argolis (Eastern Mediterranean) at 37:34 N.L. and 25:30 E.L. while serving with the Royal Navy.* (MarHis Data)

Sir James Somerville. The next day she departs, along with other Netherlands warships, to avoid an expected Japanese air raid on Colombo. That air raid actually took place on April 5, and the ships were spared by the diversion order. *Soemba* goes to Bombay for repair. The mast is shortened and additional anti-aircraft guns are installed.

On 18 April, she departs Bombay and went to the entrance of the Persian Gulf, Hormuz Strait a vital shipping lane for tankers. She is incorporated into the 'Hormuz Force', commanded by a naval captain flying his flag on the cruiser HMS *Dana*e on 22 April 1942. The commander of *Soemba* is appointed Commanding Officer Anti-Submarine Patrol Vessels, Hormuz Force. They conduct patrols and protect convoys in the Persian Gulf to Khorramshar and Bahrain. Every six weeks, the ship is off duty and rests for a week at the Naval Station Bahrain. The crew has the opportunity to play tennis, football, swim, and go to the cinema. Soemba performs this duty till 4 December, when she joins convoy to Bombay for an overhaul and repairs.

Above:
Gunboat Soemba in 1943.

On 6 January 1943, at the Naval Dockyard in Bombay, the ship is docked and the Flettner-rudder is removed completely and delivered to the shipyard. The rudder's ball and roller bearings are found to be worn. The ASDI system is improved, and an RFD system is installed. The ship also receives new camouflage colours for the Mediterranean. By February, the crew is housed at the Military Transit Camp Colaba in Bombay and goes on three-week leave in two groups, alternating between them. Ending April 1943 *Soemba* reports ready for duty again.

Soemba departs Bombay on 30 April, and goes to Aden (6-8 May) and Suez (13-14 May). Near Suez, the radar antenna becomes entangled in the cable of a barrage balloon. The cable breaks, and the balloon is carried away by the ship. A bomb attached to the balloon fails to detonate. The balloon was released in the Bitter Lakes.

Departure from Port Said on 15 May. The ship escorts a convoy of three ships with the Greek destroyer *Kondouriotis*. The next day in Alexandria, *Soemba* is assigned to Rear-Admiral Alexandria, who manages the convoy voyages. For exercises, the ship is assigned to Commodore (D) aboard destroyer tender HMS *Woolwich*. The crew's skill has deteriorated due to the long repair time and little opportunity for training. With exercises in the AA Gunnery Dome ashore, ASD exercises aboard the *Woolwich*, and exercises with a submarine, their skill is brought back up to par. Several landing exercises are held at Port Safaga, a location that bears a strong resemblance to the area where the actual landings in Sicily will take place.

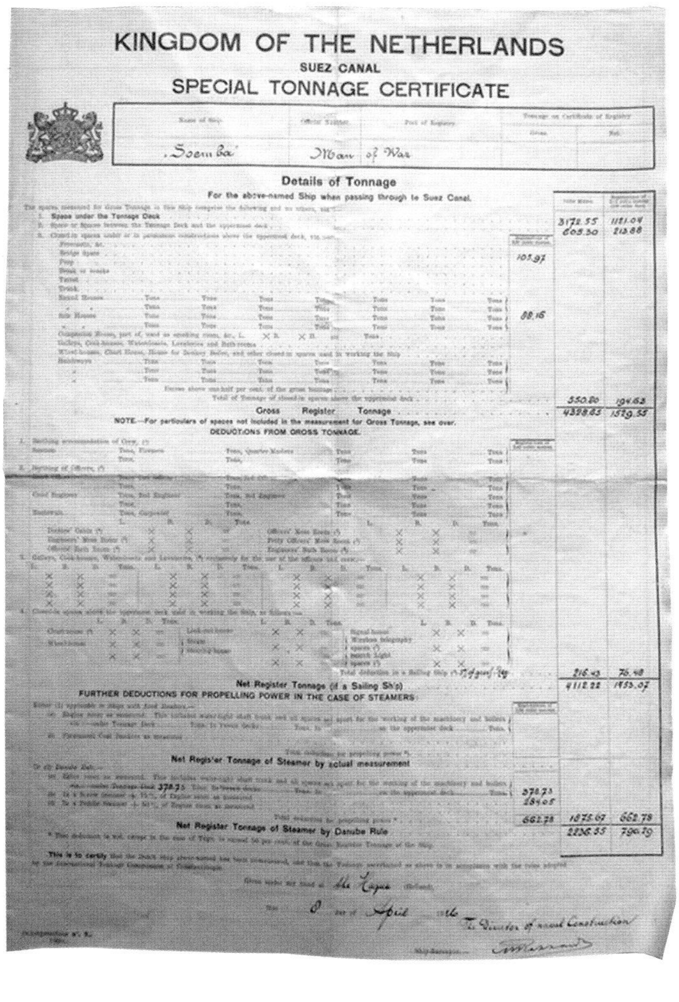

KINGDOM OF THE NETHERLANDS
SUEZ CANAL
SPECIAL TONNAGE CERTIFICATE

"Soemba" — Man of War

Details of Tonnage
For the above-named Ship when passing through the Suez Canal.

1. Space under the Tonnage Deck	3172.55	1181.04
2.	605.30	213.88
Total of Tonnage of closed-in spaces above the uppermost deck	550.80	194.63
Gross Register Tonnage	4328.65	1589.55

NOTE.—For particulars of spaces not included in the measurement for Gross Tonnage, see over.
DEDUCTIONS FROM GROSS TONNAGE.

Total deduction	216.43	76.48
Net Register Tonnage (if a Sailing Ship)	4112.22	1453.07

FURTHER DEDUCTIONS FOR PROPELLING POWER IN THE CASE OF STEAMERS

Net Register Tonnage of Steamer by actual measurement

	378.73		
	284.05		
Total deduction for propelling power	662.78	1875.07	662.78
Net Register Tonnage of Steamer by Danube Rule		2226.55	790.29

This is to certify

The Hague

8 April

The Director of naval Construction

A "Suez Canal Tonnage Certificate" is a special measurement certificate required by ships to navigate the Suez Canal. This certificate contains the ship's tonnage, calculated according to the specific measurement regulations of the Suez Canal, and is required for ships wishing to use this important shipping route.

Operations Husky, Avalanche, and Shingle

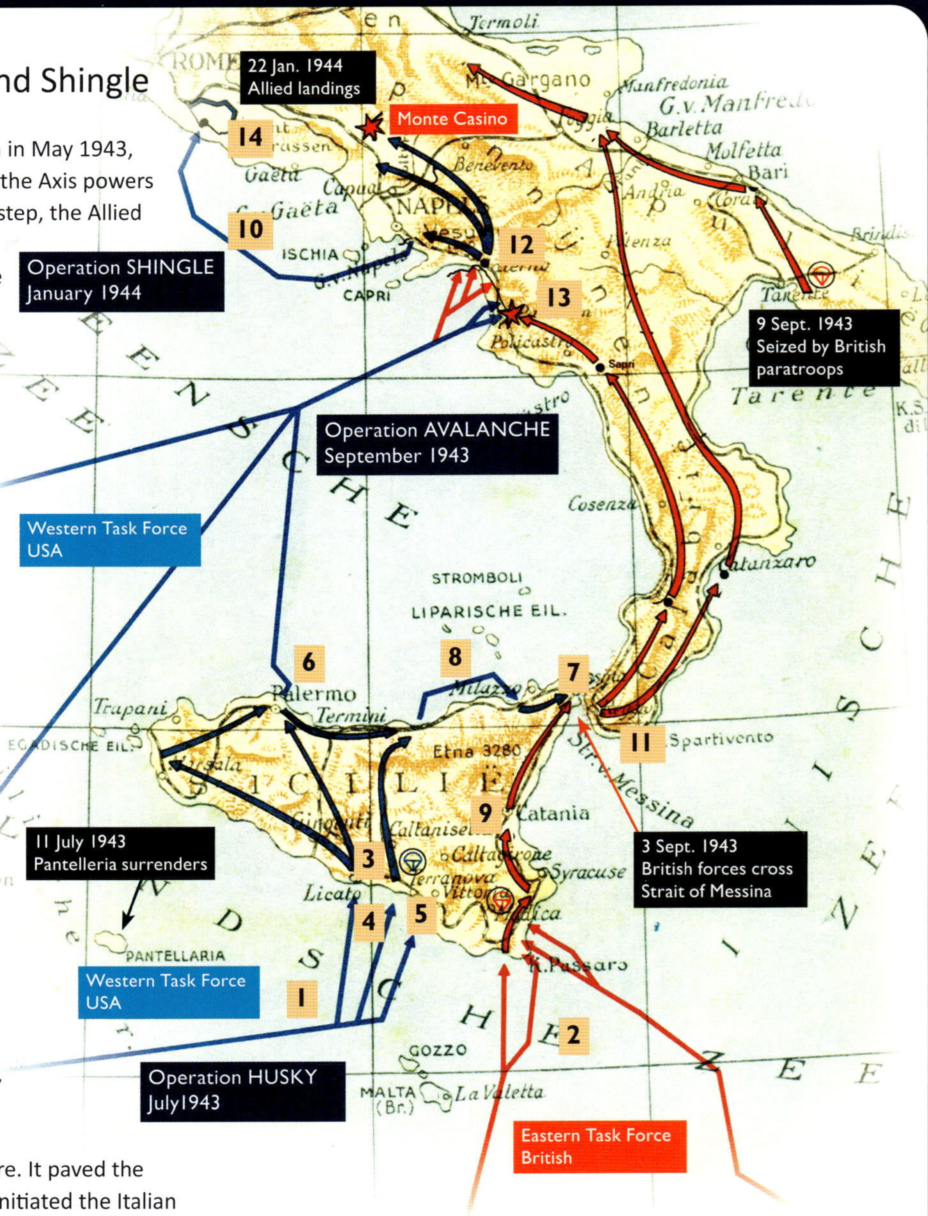

With the conclusion of the North Africa campaign in May 1943, the victorious Allies had for the first time ejected the Axis powers from an entire theatre of war. Now at Italy's doorstep, the Allied powers—led by the United States and United Kingdom—decided to attack Axis forces in Europe via Italy, rather than western Europe, due to several converging factors, including wavering Italian morale, control over strategic Mediterranean sea lanes, and the vulnerability of German supply lines along the Italian peninsula.

Operation Husky

During the Allied invasion of Sicily, also known as the Battle of Sicily Eisenhower commanded the combined operation, with the British Admiral Cunningham as the naval commander. The US Navy contingent, dubbed the Western Naval Task Force (1) and the British, the Eastern Navy Task Force (2). The Western Task Force consisted of more than 500 ships, including significant numbers of new landing craft: the 1,500-ton landing ship tank (LST), the 550-ton landing craft tank (LCT), and the 200-ton landing craft infantry (LCI). These new vessels, plus the small amphibious DUKW's (amphibious truck, duck, sloop), proved invaluable.

Despite a fierce storm on the night of 9 July, the Western Task Group successfully escorted transports carrying three Army divisions, each of about 24,000 men, to the southern coast of Sicily, landing them at Licata (3), Gela (4), and Scoglitti (5). At Gela, the cruisers covering Task Force repelled two enemy tank attacks with naval gunfire. It paved the way for the Allied invasion of mainland Italy and initiated the Italian campaign that ultimately removed Italy from the war. In the air, it was a different story. Because no escort carriers were available, US Army Air Corps planes had been charged with providing air cover for the landing. But their commander was so wedded to strategic bombing that he failed to provide sufficient forces for this vital task. Without air cover, the ships had to rely on anti-aircraft fire to repel more than 24 *Luftwaffe* attacks in 24 hours. Worse, when US transport planes arrived over the Task Force only moments after an air raid, the navy AA gunners shot down 22 of them, which were towing gliders filled with Allied paratroopers.

On the ground the 7th Army drove west then north, taking Palermo (6) on 22 July, then sped east toward Messina (7), while the British 8th Army moved up the east coast. The US movement was swifter because they took advantage of superior sea power, employing navy vessels for gunfire support and leapfrogging past Axis strong points with amphibious landings (8). Meanwhile the British became bogged down in a lengthy fight for Catania (9).

The Allied landings in Sicily eroded whatever Italian support there had been for participation in Hitler's war. In July the Italian Fascist Council stripped Mussolini from his power and incarcerated him on Ponza Island (10). His replacement immediately suggested to the Allies that Italy was willing to change sides. To support this, the Allies extended their campaign into Italy. On 3 September the 8th Army landed on the Italian tow (11), and a week later most of the US 5th Army to Salerno in **Operation Avalanche**.

The defenders were waiting for the Allies at Salerno and it turned into a pocket of death. Hoping to achieve surprise, it was decided to land without a preliminary naval bombardment. But once ashore, gunfire from ships was all that kept the Germans from driving the Allies back to sea. With the arrival of enforcements, the Germans fell back to the Gustav Line. In January 1944, the Allies did try a leap past this line to secure an open road to Rome. Code-named **Operation Shingle**, this effort was similar to Avalanche. This landing went well enough, but the troops failed to push inland. With that the Italian front turned into an uneasy stalemate.

Departure from Alexandria, on 23 May, with HMS *Delphinium*, *Burra*, the Free French aviso *La Moqueux*, and a convoy of six ships. Arriving on 27 May in Benghazi. *Soemba* continues patrolling the roadstead. Departing the same day with a convoy of twelve ships to Alexandria (31 May - 4 Jun.). Leaving Alexandria with a convoy of 19 ships. The escort consists of HMS *Antwerp*, *Maryborough*, *Ipswich*, *Burra*, and *Romeo*. The commander of the *Soemba* is Senior Officer Escort. Setting a course directly to Suez, where they arrive, on 6 June, at the Adabiya roadstead. Here *Soemba* is assigned to 'Force G' and assigned to 'Group N' under Captain Lord Ashbourne, to conduct exercises in preparation for Operation 'Husky'.

On 10 June, she exercises with several ships in the Red Sea near Safaga and a week later near Gimsah. Due to unfavourable weather, the ship steams back and forth between Gimsah and Safaga several times.
On 15 June, the ship fires 11 rounds of war ammunition as a practice for direct fire on land targets. With 11 shots, two targets are destroyed at a distance of 12,600 metres. Two days later *Soemba* returns to Suez. The ship anchors at Adabiya roadstead.

The *Soemba* conducts exercises in the Gulf of Suez, in collaboration with HMS *Carlisle* and *Antwerp* on 28 June. The ships are short of ammunition. Training is suffering, and therefore attempts are made to obtain German ammunition. Despite promises, this ammunition is not delivered. Departing on 29 June and arriving in Alexandria, the next day. The ship is further prepared for Operation Husky. The *Soemba* and its sister *Flores* are destined for the landings in sector "Bark East." The crews of the Allied ships receive training in recognising Allied aircraft. A type is announced by radio, after which the announced aircraft flies over the inner and outer harbours.

Departure from Alexandria on 3 July. The ship is assigned to the escort of convoy MWS.36, which includes the Netherlands merchant *Baarn*. It is a slow convoy bound for Sicily. The route runs along the coast of Africa to near Tripoli, then due north to east of Malta and from there to a point east of Cape Passero. After 3 days, convoy escorts attack a submarine, but it still manages to launch a torpedo. The *Sjahjehan* is torpedoed. Although the attacking submarine must have been on the convoy side of *Soemba*, no ASD contact was made. The stricken vessel is taken by tugs to Apollonia and beached there.
On 7 July, west of Benghazi, *Soemba* leaves the convoy at night and steams at maximum speed towards Tripoli. Here, the gunboat unloads oil from the tanker *Macoma*. *Soemba* has to make way for thirsty British destroyers while replenishment, resulted in a fuel supply of only 81%. When casting off the tanker, the crew of the Dutch tanker wishes *Soemba* "hearty success." Ammunition is then taken from a lighter that has come alongside. The supply is replenished to 100%, but flares are not available. The stay in port is too short to replace some of the half-armoured shells with explosive grenades.

On 8 July, 1943, departure from Tripoli. The ship joins convoy MWS.36X, which had departed Tripoli earlier at 12:00. *Soemba* joins the convoy at 15:40. Convoy MWS.36X joins up with convoy MSW.36 at 7:30 p.m., which has steamed along the North African coast through the Gulf of Serte. The *Flores* is with this convoy. Both gunboats are assigned to the invasion force on the east coast of Sicily. They are assigned the "Bark East" sector. *Soemba* leaves convoy MWS.36/36X and joins convoy MWF.36, which consists of troop transports for the landings on Sicily. This convoy includes the Dutch ships *Christiaan Huygens*, *Ruys*, and *Tegelberg*. Later, the *Flores* also joins the escort. The left column of the MWF convoy consists of HMS *Keren*, *Strathnaver*, *Otranto*, and the minelayer HMS *Brittany*. *Soemba* is placed at the head of this column. *Flores* joins the end of the left column of the convoy at 7:30 p.m.

HMS Antwerp was based at Alexandria as a convoy escort and as a troopship. In March 1943 she was converted to a fighter direction and forward operations ship, to the landings in Sicily in July 1943.

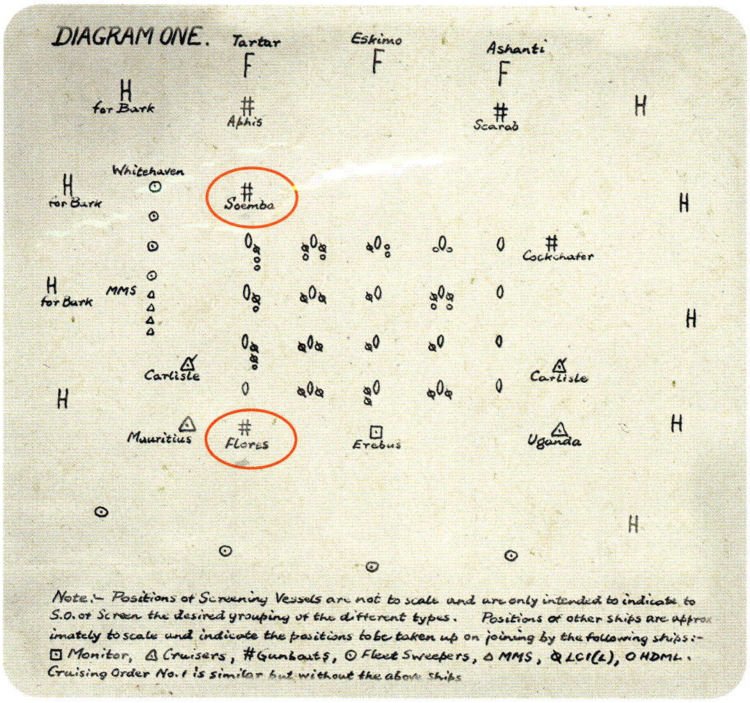

Diagram of convoy with Flores *and* Soemba.

On 10 July, the ships stopped at 00:40, while the destroyers positioned themselves on the seaward side of the transport ships to provide anti-submarine protection. The disembarkation of the landing troops began.

At 02:55, *Soemba* advanced to 6 miles to steam towards the bombardment position. The ship turned out to be 2 to 3 miles further south than planned. With the help of two sonic buoys previously deployed by the submarine HMS *Unseen*, the position was improved. The gunboat was assigned to provide supporting fire for Allied landing on the east coast of Sicily. At dawn, an enemy coastal battery opened fire on the landing craft, which were moving between the ships and the coast. The battery was located on a ridge north of Pachino and fired five guns. *Soemba* returned fire on its own initiative at 5:28 a.m. Only then did it receive a request to fire from the Senior Naval Officer Landings (SNOL), Captain Lord Ashbourne. At 6:30 a.m., the enemy battery ceased fire and *Soemba* received a telegram from the SNOL: "*Well done, Soemba.*"

At 7:18 a.m., at the request of the Forward Observation Officer (FOO), fire was opened on a machine-gun nest that was hindering the advance of the landing troops. At 10:05 a.m., *Soemba* fired on advancing enemy tanks coming along the road to Pachino. The ship fired a total of 86 shots. No more shooting took place that day, and *Soemba* anchored in Marzamemi Bay at 10:00 PM. The Netherlands merchant *Baarn* was lost that day to a heavy German air raid. Nine air raid sirens sounded throughout the day.

The following day *Soemba* relocated and signals are exchanged with *Flores*, which passes within shouting distance. During the day four air raid sirens sound.

On 13 July, both gunboats are anchored in Formiche Bay and the next day off Marzamemi. Air raid sirens sound several times. The explosives of Allies' anti-aircraft fire pose a danger. The bridge of *Soemba* is hit twice by unexploded 20 mm projectiles. No injuries occur. The "Terrible Twins" depart from Formiche Bay in the Bark South sector on 15 July, and sail via Marzamemi to Syracuse for a day's rest. During the night, the enemy carries out a very heavy air raid; no bombs fall on the harbour.

Departure from Syracuse, on 16 July, together with *Flores* to Augusta. This city had already been captured on 13 June. On 17 July, the *Soemba* is ordered to advance to Catania to bomb Misterbianco and thus support the advance of the Eighth Army, which is fighting the Germans south of Catania.

At 2:05 p.m. and 2:20 p.m., 20 salvoes are fired with indirect fire at Misterbianco. The results are not observable; the distance to the target is 17 km and is hidden behind a hill.

Landings were made in the early hours of 10 July from 2:45am on 26 main beaches spread along 170 kilometres of the southern and eastern coasts of Sicily/

Soemba *in actie op de M iddellandse Zee. Samen met een Engelse torpedobootjager wordt een rookscherm gelegd.* (National Archives)

Operation Husky was the largest amphibious invasion of the war (larger than D-Day) and still the largest of all time. It was put together in a short period of time using troops that had just finished fighting in North Africa two months prior.

Departure from Augusta on 19 July, with orders to cross the Catania plain. At 9:00 a.m., the ship is fired upon with a salvo from the shore. *Soemba* does not return fire. **At 11:35 a.m.**, a Messerschmitt Bf 109F flies over the ship at low altitude. The aircraft is pursued by a Spitfire but is shot down by the ship's anti-aircraft guns. This is a remarkable feat, as *Soemba*'s anti-aircraft guns are extremely weak compared to other ships. The success provides a significant boost to the crew's morale. They were also eager to shoot down the second aircraft, but this one managed to prove itself a friend after flying three circles around the ship. The downed pilot's parachute is retrieved and distributed among the crew. The pilot did not survive the attack.
At 12:00 a.m., the ship is again fired upon by a shore battery. By increasing the distance to shore, the risk to the ship is reduced. At 1 p.m., *Soemba* receives orders to return to Augusta. A magnetic mine has been dropped near the harbour, forcing it to be closed for half a day.

After arrival at Augusta, on 19 July, orders are received here to proceed with a convoy to Malta and load ammunition there.
At 6:00 p.m. the convoy of six landing craft and the South African minesweepers *Southern Sea* and *Protea* departs. Near Syracuse, two more landing craft and an escort vessel join the convoy. The commander of the *Soemba* is the convoy commander.

On 20 July, just before arriving in Malta, a German Junkers circles above a troop transport ship and its escort. Although the aircraft is flying quite low and clearly visible, no shots are fired at it. In Valetta (Malta), the boilers are cleaned. Since leaving Bombay, over 1,100 boiler hours have been logged. Because the boiler cleaning has to be done with the ship's own resources, it takes a week before the ship is ready to sail again.

Departure from Augusta, on 5 August. Along with the British cruiser HMS *Mauritius*, *Soemba* shells targets north of Catania, west of the town of Acireale, and a road junction near Giarre. HMS *Mauritius* comes under fire from a shore battery at Pozzallo, lays down a fog screen, and steams out of the shore at high speed. The commander of *Soemba* observes muzzle fire, but it is unknown whether this was Allied or enemy fire. The ship alters course and lays down a fog screen.
At 12:13, enemy fire falls around the ship.

The ship alters course again.
At 12:21, the commander orders to open fire on the shore battery. The order cannot be obeyed immediately because the friendly fog screen obstructs the view of the shore battery.
At 12:22 p.m. *Soemba* was hit by a direct hit on the bridge and the commander was seriously wounded. The officer of the watch, was also seriously wounded, and another sustained minor injuries upon impact. Although seriously wounded, the commander continued to consider the ship's situation. He ordered the first officer to take command and said that *"they should first attend to the other wounded."* He also asked if the ship was still under fire.
At 12:23 p.m., the course was altered on the orders of the gunnery officer, in order to remain under cover of their own fog screen until command had been reassigned. Immediately afterward, the first officer, assumed command.

HMS Mauritius, *(C80), was a Fiji-class light cruiser of the Royal Navy.*

The acting commander increases the ship's speed to full power and alters course to 90° east at 12:26 p.m., while still under enemy fire from shore. At 12:30 p.m., the last impact falls near the ship, and at 12:35 p.m., the fog clears. Course is altered to 180° east. The plan is to fire on the coastal battery. This is abandoned when it becomes clear that the wounded commander is in very poor condition and it is desirable to transport him urgently to a hospital. The ship proceeds at full speed to Augusta to bring the wounded ashore. After arrival, the commander dies during transport ashore. The other two wounded, are admitted to the 151st Light Field Ambulance.

The ship remains at anchor for repairs to the bridge, during which, in addition to a new steering wheel, a magnetic compass is also installed. The repair work is assisted by personnel from monitor HMS *Roberts*. On 9 August *Soemba* departs from Augusta for Riposto, escorted by HMS *Wheatland*, to provide support to the British 8th Army. The *Wheatland* is fired upon at 1:20 p.m. by what appears to be the same battery that fired upon the *Soemba* on the 5th. Both ships withdraw slightly and lay down a fog screen. *Soemba* then approaches the coast, and at 1:30 p.m. the gunboat opens fire, firing 29 salvos. The fifth salvo disables *Soemba*'s first gun. Two of the enemy battery's four guns are silenced. At 1:47 p.m., the *Wheatland* signals to steam beyond the battery's range and lay a smokescreen. The Commander of *Soemba* refuses and continues firing. Even when the escort again gives the order *"Go out of range," Soemba* doesn't respond. Later, the action is aborted and the ships return to Augusta.

On arrival at Augusta, the commander is told that the *Wheatland* should not have been ignored, but the Admiralty nevertheless expresses satisfaction with the action of *Soemba* at Riposto.

On 11 August *Soemba* departs for Riposto, where the ship shells the shore from 12:45 to 12:58 hours. This blows off a cover from the brake cylinder of gun III. For this reason, such a cover is borrowed from the *Flores* during the following days. Returning to Riposto same day for replenishment.

Departure from Augusta on 13 August. The *Soemba* and HMS *Scarab* bombard several positions near Riposto and Taormina in support of troops of the 8th Army. At 11:45 a.m., fire is opened. The result is good: the Forward Observation Officer reports that the intended target, a road intersection, has been hit three times. At 12:45 p.m., a bend in the coastal road is fired upon to hinder the German withdrawal. Impacts cannot be observed, and firing is ceased. The ship has fired a total of 40 salvos and returns to Augusta.

On 15 August 1943, departure from Augusta. Now with the gunboat HMS *Aphis* and escorted by HMS *Brocklesby*. Both gunboats are deployed to provide fire support to the army, but a request to fire is not received. *"We then just sailed back and forth a short distance near a cape; it later turned out that the cape had been in German hands all day,"* an officer laconically wrote in his report *"The Soemba fixes it."* Because no targets can be detected on board, the ships return to Augusta.
Two days later Sicily falls into Allied hands. The Italians and Germans have managed to transport their troops and equipment unnoticed via the Strait of Messina to mainland Italy. Shortly after *Soemba* and *Flores* receive a telegram from BDZ Vice Admiral Furstner, which reads: *"On behalf of the Royal Netherlands Navy. Offer appreciation for your work in one of the most important operations this war. Name of the Netherlands being held high and contributing to the liberation of our Kingdom."*
On 21 August, the bombardment squadron that participated in the actions off Sicily went to La Valetta.

SOEMBA and **FLORES** were nicknamed the **"Terrible Twins"** because of their activities in the campaigns off Italy and Normandy.

Gun support.

HMS Wheatland *(L 122), an escort destroyer of the Hunt (Type II) class.*

HMS Cockhafer, Aphis (pictured) and Scarab were Insect-class gunboats

On 23 August, the bombing squadron departs La Valetta. The squadron consists of the monitors HMS *Erebus* and *Roberts*, gunboats *Soemba* and *Flores* and HMS *Cockhafer*, *Aphis* and *Scarab*. While escorting a straggler from a convoy, one of the escorting fighters attacks a submarine. The depth charges explode about a mile from the *Soemba*. A moment later, the small rudder becomes stuck, and the ship is forced to stop. The main rudder is rotated 360° by hand, releasing the small rudder. The connections between the steering wheel and the small rudder are checked, but nothing is found. After arrival at Bizerta on 25 August. The ship loads oil. Maintenance and repairs are carried out. The entire steering system is also checked, but no malfunctions can be found. Two days later, the ships steam to Lac de Bizerta, where further orders are awaited.

On 1 September, the Terrible Twins are assigned to participate in the U.S. Fifth Army landing operation at Salerno: Operation Avalanche (see page 38). Both to the Western Naval Task Force under Vice Admiral K. Hewitt (USN). Departure from Bizerta, on 6 September. At 11:30 a.m., the small rudder of the *Soemba*'s Flettner rudder seized again, a situation that recurred several more times later in the afternoon, forcing the ship to be diverted to Bizerta for rudder repairs. Once returned at the Bizerta roadstead with its narrow channel and numerous wrecks, the commander decides to call for tugboat assistance. The first officer urges a speedy repair, but is told: *"Why don't you keep quiet? Don't you feel happy outside?"* Because there is no immediate tugboat assistance, the ship is not towed into port until 10 September. At Ferryville (the French repair base at Bizerta), an inspection is carried out with divers. This takes four days. The *Soemba* no longer participates in the operation at Salerno. On 16 September the ship is taken into dock for repairs. The rudder fin of the rudder has sagged and this can only be repaired while dry. She leaves dock, on 23 September, and steams to Lac de Bizerte, where the ship loads oil from a tanker and sails to to Goulet du Lac Bizerta, where the ship anchors.

Gunboat Flores.*(NIMH)*

On 28 September, *Soemba* serves as an escort between Bizerta, Palermo, Alexandria, and Gibraltar. Underway from Palermo with a convoy bound for Bizerta. Between Sicily and the North African coast, an explosion occurs (30 Sept.) under the stern of an American vessel. Because no submarines were reported or sighted, and a mine was practically ruled out, sabotage was cited as the cause of the explosion. The ship disappeared into the waves after twenty minutes; there were no casualties.

Performing escort duties in the Mediterranean. By November the crew are slowly starting to notice that the *Soemba* (built for the tropics) has no heating. On 1 January, 1944, *Soemba* was supposed to bring one ship into Valetta, but due to a heavy storm, *Soemba* had to wait out the storm. The convoy ship had to make its own way to enter. The ship's dog was swept overboard and could not be rescued. The following day arrival at Valetta where she moors alongside the *Flores*, which had arrived a day earlier. The ship undergoes repairs and maintenance. The radar and degaussing cable are repaired, and the broken glass and pottery are replaced.

On 10 January, *Soemba* departs La Valetta on her own bound for Naples, in accordance with the order of 25 December. On 16 January, *Soemba*, along with the *Flores*, is to escort a submarine to the Gulf of Salerno; the assignment is revoked before departure. They receive operational orders for a major landing exercise in the Bay of Salerno.

J-class destroyer HMS Janus *was struck by one Fritz X guided bomb dropped by a German He 111 torpedo bomber and sank. Of her crew of 183 men, only 80 survived, being rescued by HMS Laforey and smaller craft.* (Wikipedia)

In January 1944, the *"Terrible Twins"* are assigned to Peter Force to land the British First Division and are not ordered to sail to participate in Operation Shingle: the British-American troop landing at Anzio. The British Naval Liaison Officer (BNLO) of *Soemba* is sent ashore because it is suspected that both gunboats have been "forgotten" by the Americans. Officially, the BNLO is informed that they were originally intended to act as communication stations near Ischia, but that an US signal brigade has been stationed on the island itself. The Americans therefore believe the ships are unnecessary. However, the British Flag Officer, Naples, disagrees and sends a Bombardment Liaison Officer (BLO) aboard and assigns the ships to the British sector west of Anzio. The BLO has the operational orders and graduated maps of the action.
The next day, *Soemba* anchors three miles from the beach near the headquarters ship of the British landings, HMS *Bulolo*. Because of fire from the shore and shells falling nearby, the *Soemba* weighs anchor and bobs back and forth for an hour and a half. Then the ship anchors again.

On 23 January, between 01:12 and 02:00, an E-boat alarm is sounded. While firing is heard, no enemy ships are seen. *Soemba* has no task assigned to Operation Shingle. The Senior Officer Bombardment Group signals: *"Sorry, no target for you."*. The landing operation appears to proceed without significant opposition. Throughout the day, there is a continuous air raid siren, but most of the attacks take place in a different sector. The ship fires -without result- on a Junker 88. During this attack, destroyer HMS *Janus* receives a direct hit and sinks with a large part of its crew. Later, another destroyer HMS *Jervis* is severely damaged by a rocket. The admiral sends the cruisers HMS *Orion* and *Spartan* back to Naples to avoid exposing them unnecessarily to air attack. This places the bombardment squadron under the command of USS *Brooklyn*. This group consists of the *Soemba*, *Flores*, HMS *Laforey*, and HMS *Loyal*. The next day, the ship is assigned Forward Observation Officer 3 to direct the fire.

At 13:57, *Soemba* opens fire on tanks housed in a factory. Using the Anzio lighthouse as an auxiliary aiming point, *Soemba* fires, the final salvo consists of three shots. Observation is difficult, and the fire is poor. Later, she intercepts an urgent request from the Senior Officer P Force to HMS *Laforey* to shell heavy shore

batteries at Anzio. On its own initiative, *Soemba* steams closer to shore to fire on the aforementioned batteries. The ship receives orders from HMS *Laforey* to shell a road intersection, without shore-based observation support. At 3:53 p.m., the ship, while stopped, opens indirect fire on the road intersection. Eighteen shots are fired; the result is unknown. During the night, *Soemba* keeps up and down the shore with the *Flores*, *Laforey*, and *Loyal*. The ships have misted the position to protect it.

15 cm shell.

On 24 January , HMS *Orion* returns to the battlefield and takes command. The *Orion* urges *Soemba*, which is sailing close to shore, to move further out. Because *Soemba*'s Forward Observation Officer (FOO) has assigned a target, the ship is allowed to remain in position. At 11:28 a.m., the FOO requests fire on an enemy battery. Fire is opened at 11:44 a.m. but has to be stopped after 11 shots because the FOO, due to poor visibility, can no longer properly direct the fire. The FOO then continues to postpone fire due to poor visibility. Finally, at 3:25 p.m., he reports that the target has been captured by British forces and that further targets are out of range of the *Soemba*'s guns. The *Soemba* relays this message to the other ships. In the afternoon, the ships are fired upon from shore, and the *Soemba* lays a smoke and fog screen. Due to an approaching storm, *Soemba* steams out of shore at 4:45 p.m. During the night, the *Flores* is lost sight of. The target is very inaccurate, as the shore is not visible.

On 26 January, contact with *Flores* is re-established. Because the weather is deteriorating, both ships are deemed unsuitable for bombardments. They lack stabilised sights and therefore cannot deliver accurate fire in bad weather. Nevertheless, an FOO observer is assigned to them.

HMS Orion *was a* Leander*-class light cruiser which served with distinction in the Royal Navy during the Second World War. She received 13 battle honours.*

At 5:30 p.m., they receive orders to go out and spend the night in a back-and-forth position. The next day she steams to the bombing range; while the *Flores* heads for Naples with storm damage. *Soemba* was also signalled to proceed to Naples, but this message was not received.
At 1:11 p.m., the halted *Soemba* fires indirect fire at a German headquarters. The first and second salvos land in the sea. The shells then hit the shore, but after 22 rounds, the accuracy of the fire cannot be determined. The FOO apparently assigned the wrong target.

At 1:36 p.m., the FOO assigns a new target. At that moment, one of the ships is being fired upon by a shore battery, which causes a delay. The ships must be ready for fog. At 2:18 p.m., the FOO assigns the target again, and within five minutes, *Soemba* opens fire on enemy positions. After two salvos, firing ceased. At 3:30 PM, *Soemba* departed for a new bombardment position, which it reached within fifteen minutes.

At 3:55 PM, shelling resumed. After 74 shots, firing ceased. According to the observer, the fire was good. *Soemba* steamed out again for the night. The next day the gunboat is recalled to Naples.

On 1 February, *Soemba* steams to the rendezvous, which is reached at 7:45 a.m. They have to pass through a swept channel to the Formia front. There is no visibility on shore, the weather is bad, and *Soemba* loses sight of *Orion* for some time. The cruiser enters the channel with the help of a radar buoy, but this is impossible for *Soemba* with its inadequate radar. *Orion* directs *Soemba* to position 3 miles further inland, opposite the radar buoy. Due to the bad weather, aircraft observation was not possible, and after receiving a report of a sighted enemy two-man torpedo, the ships headed for the Gulf of Pozzuoli to spend the night there.

The next day, the ships are back at the radar buoy at 8:00 a.m. *Soemba* is assigned a radar buoy, with which contact is made at 8:28 a.m. Two minutes later, the ship receives a list of seven targets: the same batteries that had been designated as targets the previous day. The targets are for the cruiser *Orion* and *Soemba*. The gunboat places a marker buoy at the bombardment position.

At 10:16 a.m., two targets are assigned. *Soemba* opened fire on an battery at the foot of Monte Campese. The bombardment, with a stopped ship and indirect fire at a range of 13.5 km, was insufficiently successful, despite being guided by a spotter plane. After 21 shots, the fire was shifted to another target. The pilot of the plotter plane reported the firing as "fairly effective." After 27 shots, the firing ceased.

At 12:31 p.m., *Soemba* opened fire on an enemy battery. While engaging, the ship came under accurate fire, and after 12 salvos, *Soemba* also withdrew, laying a fog curtain at the radar buoy. HMS *Orion* did not abandon *Soemba* but remained in the immediate vicinity, heading for a safe position. After this, the ships ceased to move back and forth. The battery that had been shelling *Soemba* was under fire from Allied field artillery. The ships steamed back into the mine-free canal and opened fire on an enemy battery, firing 33 rounds.

At 4:12 p.m. , the gunboat fired at a road intersection 17 km away. The *Orion* returned to Naples, while *Soemba* remained at the bombing range. During the night, she returns to the swept canal and carries out shelling the next day. She fires 94 shots. Gun II is put out of action due to the collapse of its cradle. At 5:45 a.m. fire ceases. The gun malfunctions, causing several casualties. Word is received that *Soemba* will be relieved by *Flores*, but *Flores* cannot immediately locate *Soemba*.

At 11:53, she fired at an enemy battery. The ship was stopped, and the target was approximately 14 km away. The fire was successful. Observation was provided by a plotter plane and the FOO. A total of 30 rounds were fired. Shortly after, *Soemba* effectively fired at another battery, also approximately 14 km away. After 34 shots, she had 10% of her ammunition remaining. Due to the low ammunition supply, only one target could be engaged. In the afternoon, fire was directed at a third battery. But was broken off after 10 shots because the aircraft spotter observed the wrong target. Due to the depleted ammunition supply (38 shells remaining) and the disabled gun II, no further shelling is carried out, and *Soemba* requests permission to leave the bombardment range. The Flag Officer Naples is signakled that the guns are running out of ammo and *Soemba* is ordered to return to Naples.

Departure from Naples roadstead on 4 February, en route to Malta. While steaming out of the bay, *Soemba* encounters twin-sister *Flores*, with which messages are being delivered by line. The voyage to Malta (6 - 21 Feb) is made in a large convoy. On 23 February, both gunboats are ordered to depart for England to receive guns from the ageing cruiser *Sumatra*.

In March 1944, Queen Wilhelmina visited Portsmouth to inspect Dutch naval ships, specifically the RNN gunboats Flores *and* Soemba, *and to present medals. She was accompanied by Prince Bernhard and also met with the Commander-in-Chief, Admiral Sir Charles J.C. Little.*

Departure from Valetta on 25 February as escort in a small convoy to Gibraltar. After replenishment in Gibraltar, both gunboats are assigned to escort a 30 ships convoy. On 9 May, one of the escorting ships, Flower-class corvette HMS *Asphodel* (K56), is torpedoed. The following day, both gunboats and an English net layer detach from the convoy and make a beeline for the Irish Sea, because they are not allowed to sail along the English coast at night, due to the danger of E-boats. On 12 March, *Soemba* is on submarine alert. Although it is a school of fish, a depth charge is dropped anyway: *"We hadn't eaten fresh fish for three months."* A boat is lowered to bring in the catch. The fishermen's flag is hoisted, and *Soemba* quickly steams after the other two ships. The fishermen's flag causes considerable confusion among the signallers of the other two ships. The ships are mistaken for German fighters by a British aircraft, after which a group of British bombers appears on the scene. However, using Dutch flags spread out on the ship, the pilots are able to determine the identity of the ships in time. Shortly after, they anchor in the roadstead of Portsmouth.

On 22 March, 1944, Queen Wilhelmina and Prince Bernhard visit the ship. Thirty-six Bronze Crosses are awarded to crew-members of both gunboats. By the end of the month, *Soemba* is repaired at the Portsmouth Dockyard. The guns are replaced by guns from the cruiser *Sumatra*. The radar is replaced and expanded. Engine and boiler repairs are also carried out, and the Flettner rudder is removed for the repair of, among other things, the bearings and gaskets. Halfway through April, she is taken into dock. About a month later, she starts trials of new equipment and the commander gets a brief overview of the upcoming operations in Normandy for the first time and learns what the *Soemba* 's task will be. After much insistence, a VHF radio type 86M is brought on board. Experiences during the landings in Sicily had made it clear that communication with

Flores *en route to Normandy in June 1944*

the radio system in use at the time was inadequate. The British refuse to install the device due to a lack of time. *Soemba* goes to Yarmouth.

On 24 May, she went to Plymouth. From this port, the ship will depart to participate in the invasion of Normandy: Operation Neptune. The Commander of Force Utah, holds a meeting here, about the upcoming operations on the coast of Normandy. A letter from the commander of *Soemba* reveals that the ship cannot receive the necessary communication equipment in time to maintain contact with the fire control devices ashore. The ship will, however, be equipped with radio equipment to contact observation aircraft. Incidentally, the Rear Admiral appears to have been unaware of this and believes that *Soemba* does not have any fire control. This may be the reason why the Americans expected so little from *Soemba* 's contribution. Furthermore, the commander of *Soemba* learns that his ship has been assigned to a slow convoy, which is only moving 5 miles while the rest of the bombardment group is steaming ahead at 14 miles. The commander therefore requests to be placed in that group in order to arrive at the designated bombing position in time. All this gives the impression that the Americans have no significant role in mind for *Soemba* . The Americans are also later unwilling to support the American Rangers at Pointe du Hoc. The task during the initial landings thus becomes only firing on two targets: from 40 minutes before the start of the landings until 2 hours afterward. It remains strange that a ship's commander has to peddle his ship in this way.

Operation Neptune

During this operation, which lasted from 6 to 24 June, more than 4,000 Allied vessels landed 714,000 men, more than 111,000 vehicles, and more than 250,000 tons of supplies in France.

Normandy was selected over the Pas de Calais (1) because the Allies needed the port facilities at Cherbourg (2) and Le Havre (3). Until those ports were secured, however, the Allies would have to rely on supplies arriving over the beach, and to facilitate that effort, British planners devised innovative floating docks called 'Mulberries' and an offshore breakwater of sunken hulks to create a temporary harbour.
The weather provided some last-minute drama. Originally planned for May, it was finally set for 5 June. But on 4 June the channel blew up into a gale, forcing to postpone the attack.

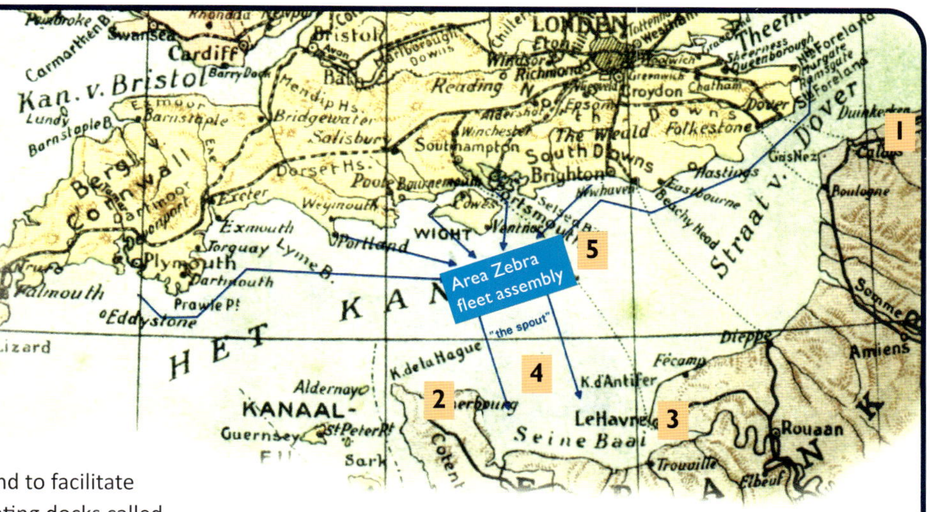

Minesweepers cleared a channel (the Spout) through an Axis mine field (4) and marked the channel with buoys, while ships from nearly every port in Southern England assembled at Area Zebra (5) for the push south. The Germans assuming the poor weather would keep the Allies in port were unaware the Allied armada had sailed.

On 3 June, departure from Plymouth; the ship anchors in the Sound. **At 1:15** p.m., *Soemba* weighs anchor and steams behind USS *Barton* (DD-722), the convoy's command ship, to Weymouth, where convoy U2A is being formed. The *'Terrible Twins'* are the only Netherlands warships providing fire support. *Soemba* is assigned to the Western Naval Task Force (USA) in the Utah sector; the *Flores* to the Eastern Naval Task Force (British). A third Netherlands warship, the outdated cruiser *Sumatra*, participates in the operation as a blockship and will be sunk on 9 June as part of the artificial harbour 'Gooseberry 5.'

6 June 1944, D-day.

Faster convoys must be given priority more than once. As a result, our own convoy suffers considerable delays. *Soemba* enters the swept channel independently at 3:55 a.m. and arrives at the bombing position at 6:10 a.m. The ship, along with the British cruisers *Enterprise* (D52) and *Hawkins*, forms Group II of the bombardment ships, which provide fire support in the Utah sector.

While underway, *Soemba* opens indirect fire on a fortified position on shore (Objective 30). The target is at 7,600 m. It is an infantry position in the built-up area of Grandcamp-les-Bains, consisting of two pillboxes and one light gun on the harbour head. Three guns are reported in the vicinity. *Soemba* fires 15 shells. At 6:17 a.m., firing ceases. The result remains unknown, but impacts are reported on houses in the area. At **6:20 a.m.**, the ship, while

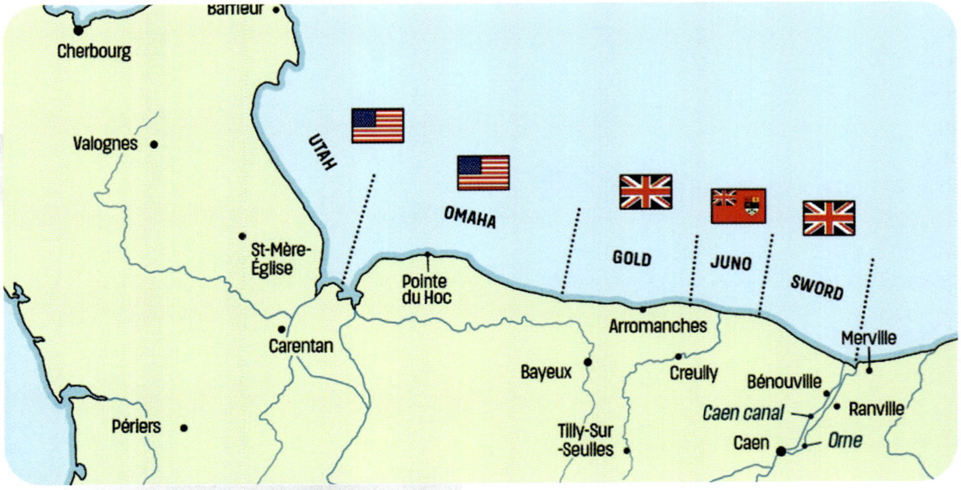

Operation Neptune was the naval component of Operation Overlord: the codename for the Allied invasion of north-west Europe. Nearly 7,000 vessels took part in the landings on the five Normandy beaches.

Empty shell cases being stores on deck of Flores.)NIMH)

A roundabout is held at Portland because D-Day has been postponed until June 6 due to bad weather. The convoy is brought to Weymouth Bay while *Soemba* is towing *LCT 2226*, which has mechanical damage. The ships anchor on 5 June at 1:45 a.m. at Portland Harbour and depart at 4:45 a.m. Initially, *Soemba* is positioned 3,000 metres behind the convoy, but at 12:10 p.m. the ship is stationed 500 metres in front of the convoy, probably as a result of a written protest from *Soemba's* Commander. The convoy leader's orders are not always clear, but they still manage to approach the swept channel successfully. A floating mine is fired upon, but without result.

underway, opens direct fire on a fortified infantry position (Objective 32) on the west side of Grandcamp-les-Bains, consisting of three pillboxes, one 3.7 cm anti-tank gun, one 4.7 cm anti-tank gun, and a shelter. Two 15 cm guns are also reported in the area. The target is 9,000 meters away. Here too, 15 rounds are fired. Hits on houses are reported. The fire ceases after two minutes. A few minutes later, fire is directed at the gun on the harbour head of target 30. Range: 7,800 metres. Twelve rounds are fired until 06:28. Then, firing ceases because smoke and dust from other bombardments obscure the target. The outcome of the bombardment remains unknown. So far, no activity has been observed at the targets, and no return fire is returned. At 06:31, muzzle fire from an anti-aircraft battery is observed on the east side of Pointe du Hoc. A minute later, *Soemba*, while underway, fires directly at the battery, which is over 8,000 meters away.

At 06:35, the ship ceases fire after firing 15 rounds. Previously, deviations of +/- 600 metres had been observed during the attacks. Now that these have been corrected, the fire is considerably better. Aboard *Soemba*, there is little to no information about the landings in the Omaha sector. Given the heavy fighting taking place there, *Soemba* would like to assist. The ship is located close to the dividing line between 'Utah' and 'Omaha.' But acting independently is, of course, out of the question. As far as can be observed, the landings on Utah Beach are progressing well. In the Utah sector, targets 30 and 32 can still be fired upon a few times:

Flores *on Operation Neptune*

(National Archives)

At 07:03, direct fire was directed at target 32, located at 9,400 m. After 33 rounds, firing ceased. The result was unknown. Thirty minutes later, 18 rounds were fired at target 30, located at 8,100 m. The ship stopped and returned direct fire. Again, the result was unknown. While stationary, *Soemba* fired 15 rounds directly at target 32 at a range of 9,400 m. Due to poor observation, no result could be determined here either.

At 08:21, the ship fired 14 rounds at target 32. The ship stopped and provided direct fire over 9,500 m. Shortly after the ship again opened fire on target 30. While stopped, she provided direct fire at 8,800 m. This time, 18 rounds were fired. The outcome of these shelling remained unknown. *Soemba* remained in her bombardment position for the rest of the day. Although the other ships were constantly in action, *Soemba* was not requested to provide artillery support. The ship was completely uninformed about the situation ashore. Thus, remained useless at the scene. Cooperation with the Americans was difficult. They preferred to use radiotelephony for shelling, for which *Soemba* did not have the necessary equipment. The commander regularly volunteered his ship for operations, as they felt they could make a useful contribution. However, due to the aforementioned causes and the sometimes confusing situations on the ground, the Americans are likely unable to properly command the ship. *Soemba* chooses an anchorage on its own initiative at night. The next day, *Soemba* returns to the bombardment position from the previous day. An American destroyer fires on a casemate on the beach. During a lull in shelling, an American jeep emerges from behind the target. A moment later, the shore signal post reports that the casemate has been in American hands since the previous day!

Shelling continues on the Utah sector, but the bombardment group of *Soemba* have nothing to do. The British call it a day and steam over to the Omaha sector. After begging for orders again but receiving no response, *Soemba* follows the British. That evening, the ship anchors in the Utah sector.

On 8 June, at 00:10, enemy aircraft are heard. *Soemba* fires on a plane flying over the ship. The ship is assigned a bombing position and observer, but due to the previously mentioned radio problems, *Soemba* cannot contact them. However, the new bombing position brings the ship closer to the coast in the western part of the landing zone. Throughout the day, the reports remain very confused. Muzzle fire from batteries is observed in the direction of Quineville. The ships in this sector do not respond, and *Soemba* requests from the USS *Tuscaloosa* (CA-37) The requested permission is granted. However, at 14:15, the ship itself comes under fire and the anchor has to be weighed. Because the fire is accurate, the ship is forced to astern at full power, thus pulling the anchor out of the ground. A subsequent strike falls precisely on the spot the ship has just left. While withdrawing, two mines explode (activated by minesweepers), but there is no damage. *Soemba* resumes position at the bombardment site but does not anchor. The commander wants to fire on anything that strikes him as suspicious and requests

Above and below: *Gunboat Flores at D-day.* (National Archives and NIMH)

information from the *Tuscaloosa* about its own positions on the beach around Quineville. Neither the *Tuscaloosa* nor the destroyer *Laffey*, which was later contacted, is capable of doing so. Although the commander decides to fire on all muzzle fire, this does not happen. A barrage balloon becomes entangled in the mast. While attempting to free it, the *Soemba* receives a signal from the Senior Officer Bombardment Group: *"Stop that running around; remember there are still mines."*

Meanwhile, an American ship has "stolen" *Soemba*'s target. The orders for the night are unclear, and since the *Tuscaloosa* doesn't respond to requests for clarification, *Soemba* anchors in the same place as the previous night.

On 9 June, at 12:21 a.m., orders are received to remain at the bombing range. An order for destroyers appears to have also been intended for the gunboat. Given the mine danger and the fact that no bombing had taken place during the night so far, it was decided to remain at the anchorage. *Soemba* is assigned a Shore Fire Control Party (SFCP). With the asdic, splashes are heard in the water, and at 1,000 meters an aircraft bomb explodes.

At 07:00, *Soemba* departs for the bombing position. It again turns out that the gunboat does not have the correct equipment on board, and contact with the SFCP is therefore not established. *Soemba* remains idle all day. Contact with the other ships is poor. During air raids, the merchant ships fire so wildly that the commander decides it would be wiser to choose an anchorage a bit further away. Time is passed by patching and sewing.

The next day they weighed anchor at 6:00 a.m. to take up a new berth. *Soemba* was assigned a new SFCP, but again, no contact could be made.

At 7:35 p.m., muzzle fire was observed, and *Soemba* opened fire on the shore battery at Quinville at 7:38 p.m. The ship fired 30 shells, with unknown results. The ship's rangefinder proved to be malfunctioning, giving a deviation of more than 1,500 m. The ship anchored during the night.

On 11 June, at 6:30 a.m., anchor is weighed to fire 20 shots, as requested, at targets 88 and 90, two fortified positions on the coast. *Soemba* while stopped, opens direct fire on both targets, which are located at 6,000 m. Forty-six shots are fired. Seven minutes later, firing ceases. The outcome remains unknown. The fire was intended to support American troops on thew-estern flank of the landing front. Muzzle fire is observed near Quinville, and *Soemba* requests and receives (!) permission to open fire on it.

At 11:54 a.m., *Soemba*, while stopped, begins direct fire on the enemy battery, located at 9,500 m. The ship fired 39 rounds. This time, there were reasonable results, and the enemy battery silenced.

At 7:52 p.m., another enemy battery 12 miles away was fired at 30 shots. After 8 minutes, the shelling ceased. The enemy battery ceased firing. About an hour later, *Soemba* opened fire on an enemy battery of 3 guns, firing 23 shells. The enemy battery, 17 miles away, ceased fire. During the night, the ship anchored at the bombardment position.

On 12 June, artillery fire was observed near Quinville. *Soemba* was assigned a bombing position and an SFCP.

At 13:45 hrs, muzzle fire from three guns of a shore battery is observed. *Soemba* opens fire on the battery, which is almost 19 km away. After 21 rounds, firing is stopped, as there are strong doubts about the outcome of firing without an observer. The gunboat has only 29% of its ammunition left.

At 14:43 hrs, at the request of a US destroyer, direct fire is given without observation at a road intersection 12 km away. After 62 rounds, firing is stopped. The ammunition supply had now been reduced to 10%. The result of the bombardment was unknown. The bombardment squadron commander was informed at 3:40 p.m. of the reduced ammunition supply, and the Force U(tah) commander-gave the order to send *Soemba* to Plymouth with a fast convoy. However, when the

USS Bayfield (APA-33), flagship for the Utah Beach landings, loading landing craft on D-Day, 6 June 1944.
(Official U.S. Navy Photograph, now in the collections of the National Archives.)

ship reported to HMS *Ceres*, a convoy had just departed and *Soemba* was ordered to steam individually to Plymouth to load 500 shells. Although the radar was broken and the asdic unreliable, no objections were raised, and the voyage was continued. The ship nearly collided with the anti-submarine yacht *Hiniesta*. During the night, flares, parachute flares, and gunfire were observed, and several smaller convoys were also sighted. However, *Soemba* reaches the other side without any problems.

Arriving in Plymouth on 13 June, and US ammunition (due to a lack of Dutch ammunition) is loaded. A large number of shells turn out not to fit, and a new supply has to be ordered. Some shells also appear to have shorter lids, requiring the ammunition lifts to be slightly modified. *Soemba* receives two departure notices (before June 15, then before June 14) to depart with a convoy, but ultimately the ship remains at the buoy in Plymouth with the order to be ready for steam within four hours. A Very High Frequency, VHF 609, with two frequencies, is installed on board, and the asdic and radar are also repaired.

On 25 June, during the night, USS *Bayfield* (APA-33) arrives in Plymouth. The liaison officer from *Soemba* is sent aboard. He is informed that the Western Task Force had been changed. *Soemba* is ordered to depart as an assigned escort ship for a convoy for the British sector near Normandy to relieve the *Flores*.

On 28 June, at 8:30 p.m., departure from Plymouth escorting the US steamship *John Henry* to the lightship Eddistone Rock. Two hours later, the first leg of convoy ECM.17 is joined at Eddistone Rock. The next day, the second part of the convoy joins the Isle of Wight.

At 12:10 p.m., *Soemba* leaves the convoy to steam as quickly as possible to the 'Sword' sector to report to HMS *Belfast*. She arrives at 'Sword', just as this sector is being shelled by an enemy battery. The ship drops anchor, and a meeting takes place between the commanders of *Soemba* and the *Flores*, while books and orders are taken over. During the night, there is enemy air activity, probably only for dropping mines.

On 30 June, no Forward Observer Bombardment (FOB) is assigned for this day. *Soemba* lays a heavy fog screen during the day near the French town of Ouistreham (at the mouth of the Marne). Muzzle fire is seen in the hills around Cabourg,but due to the great distance, no return fire is fired.

At 3:00 p.m., the ships are fired upon, as well as at 9:50 p.m., but no muzzle fire is observed. In the night, there was a continuous air raid siren, but no bombs fell near *Soemba*.

On 2 July, during the night, there was a continuous air raid siren. In the morning indirect fire was opened on a troop concentration at Bavent. The distance was 16,300 m. Eight shots were fired. After a minute, the firing, which was carried out without observation, was stopped. The result was unknown. In the afternoon, *Soemba*, fired indirect fire on Bavent. After 26 shots, the firing ceased. The observer reported: *"Good shooting."*

At 7:45 p.m., the roadstead was shelled by a battery of 4 to 6 guns. In return, shells landed at a distance of 700 m from *Soemba*.

On 3 July, *Soemba* opened fire on Le Hoin without being sighted. In one minute, 11 shots were fired. The target was 18 km away. The result of the indirect fire was unknown.

At 4:34 PM, a bombardment of targets near Robehomme begins, at a distance of over 18 km. Twenty-two shots are fired in 10 minutes. The indirect fire proves effective.

At 5:12 p.m., *Soemba* directs indirect fire at several targets at a distance of approximately 16 km. After 29 shots, firing ceases

On 4 July, *Soemba* steams closer to the coast at Ouistreham to fire on a target east of Orne near St. Richer. About 30 minutes later, indirect fire is opened on the village of Bavent, which is 15 km away. After 24 shots, firing ceased. The observer had difficulty spotting and reported *"Not seen"* for most salvoes.

At 4:56 p.m., *Soemba* gives indirect fire on a battery 16 km away. 39 shots were fired. Two hits were observed on the target. In the evening, on orders from HMS *Belfast*, the bombardment ships weighed anchor, steamed to the Juno sector, and anchored there again The following day *Soemba* weighs anchor and returns to the bombing range. There indirect fire is opened on a target near Bavent, 17 km away. Forty-two shots are fired, after which firing ceases. Only guns I and II are fired. Although the shooting is effective, but half of the shells fail to explode...

At 7:12 p.m., bombardment begins on the battery near Dives-sur-Mer, at a distance of 18 km. Twenty shells are fired, and at 7:36 p.m. the last shot is fired. The bombardment is successful. She selects a new target: an enemy observation post 14 km away. Laster, indirect fire begins, and after six shots, fire ceases. The target is not hit and *Soemba* weighs anchor again.

On 6 July, the gunboat steams back to the bombing position. But two warnings are issued before noon about the threat of small German submarines and two-man torpedoes. It is decided to keep *Soemba* on standby in the 'Juno' sector, in the late evening the ship departs.

On 8 July, the Polish cruiser ORP *Dragon* (ex-HMS *Dragon*) was torpedoed by a German Marder manned torpedo. The cruiser was 700 meters from *Soemba*, which until the previous evening had been the outermost ship of the bombardment ships. *Soemba* dropped some depth charges as a precaution. Shortly after she departed for the bombardment range.

At 07:20 a.m.,,*Soemba* opened fire on what appeared to be a small submarine, but it turned out to be the wreck of a MTB, which was just barely sticking out of the water. In the afternoon , *Soemba* opened indirect fire on several targets near Bavent. She fired 63 rounds with effective results. After 30 minutes the shooting stopped. While steaming to a new bombardment position, an enemy battery opened fire. *Soemba* was unable to anchor, and the planned bombardment was abandoned. The FOB also cancelled the bombardment because enemy rocket fire was hitting close to it. The gunboat returned to the 'Juno' sector.

AA gun of Flores *in June 1944*
(National Archives)

In 1944, HMS Belfast played a crucial role in the Normandy Landings (D-Day), serving as the flagship for Bombardment Force E, leading the initial shore bombardment for Gold and Juno beaches on June 6th, and providing artillery support for troops inland for over a month, marking the last time her guns fired in WWII.

On 9 July, around 3:00 a.m., a submarine alarm was sounded. Several landing craft circled the bombing ships and dropped depth charges. *Soemba* returned to the bombing range. In the afternoon a target was assigned, but due to the position of Allied troops, it could not be fired upon. Shortly after. she opened indirect fire on a target south of Franceville. The distance to the target was over 14 km. Seventeen shots were fired and the shelling, which was ineffective. Observation of the target was difficult. After an hour *Soemba* begins bombarding a target near Gonneville-sur-Merville, 14 km away. The ship fires 15 rounds. The result of the bombardment is unknown. Another bombardment is carried out on a target near Home-sur-Mer. The target is over 14 km away. In 19 minutes, 54 rounds are fired. The result is effective. In the afternoon, indirect fire is brought to bear on a battery, but without success. After 9 rounds, firing ceases because the spotter cannot detect any impacts. *Soemba* steams back to the 'Juno' sector.

On 10 July, numerous mines explode at the anchorage, probably due to the swell. There are also numerous air raid sirens. She departs for the bombing range and makes contact with the FOB. Shots are fired at a fortified position at Pointe Homme. After 14 shots, no results are observed. The shelling takes place under difficult conditions: the wind makes it necessary to weigh anchor so that three guns can fire. However, due to the mine hazard, it is difficult to keep the ship in the correct position with the propeller.
Enemy aircraft drop magnetic mines near the gunboat, twelve of which detonate. The ship is not in danger. Receiving a telegram ordering all shipping to stop due to the danger of mines. In the evening *Soemba* fired 47 shells at various targets near Bavent. The results were good. Due to the limited ammunition supply (5 charges per gun), firing was halted and she remained at the bombing range overnight.

The next day, 11 July, about, 2½ miles from *Soemba*, the *LCG 831* struck a mine and sank. Throughout the day, the mine-sweepers in the area were successfully active. *Soemba* received orders to steam to Portsmouth. She sails through the swept channel, passes the lightship *Juno* and proceeds to Plymouth along the designated route. Along the way, several "flying bombs" are seen flying toward Plymouth.

Gunboat Flores near the Normandy beaches in 1944. (National Archives)

On 12 July, 1944, *Soemba* arrives at Portsmouth. The Chief of the Naval Staff in London has meanwhile proposed to the British Admiralty that *Soemba*, due to its advanced age, no longer be equipped with British guns, and that it be laid up. Due to a lack of space in Portsmouth, the ship is therefore directed to Harwich. Departure from Portsmouth with a convoy to Harwich on 20 July. En route, the crew can see the defences against the flying bombs.

On 21 July 1944, arrival at Harwich (21 Jul. - 8 Aug.). She is laid up with a core crew. The ship is moored. Arrival at London on 8 August. The ship moors in Shadwell Basin alongside the Netherlands Passenger-/cargo vessel *Oranje Nassau*, and is laid up on 26 August 1944. The main battery is no longer usable for war purposes due to its poor condition. The ship remains in use for training purposes and has a sailing readiness period of one week.

On 25 September, 1944, she is commissioned as a separate unit and is designated as a training and depot ship. On board, training for the newly arrived navy personnel begins.
The Dutch naval staff requests the British Admiralty to equip *Soemba* as a radar training ship. The request is granted.

Rounds fired Operation Neptune 1944			
6-jun	155	2-jul	26
10-jun	30	3-jul	33
11-jun	138	4-jul	63
12-jun	83	5-jul	68
		8-jul	63
		9-jul	95
		10-jul	61
Total	406		409

On 8 May, 1945, *Soemba* departs for Chatham Dockyard via Sheerness. A few days later, the three 15 cm guns are landed. The main battery and the fire control system are intended for artillery training. Only the Oerlikons remain on board. She departs Chatham Dockyard on 19 May, and goes to Grangemouth were she is converted into a radar training ship. The ship's name signal becomes HX 1 on 9 August.

In April 1946, oil is being loaded at Grangemouth roadstead. However, one of the engineers forgets to loosen a relief valve. The rising pressure caused a hull to rupture, destroying a recently installed radar set and smearing the necessary equipment with oil. After three days, the ship was ready to sail again.

On 2 May, she departs from Grangemouth for sea trials in the Firth of Forth. Full-speed trials are conducted, the ship steams over the range, the compasses are compensated, and the ground tackle is tested. The following days the final repairs are carried out at Chatham Dockyard. Victuals and inventory goods are loaded. Passengers also come on board: eight women and four children. Two prams also make the upcoming journey. Then the ship is towed to the oil jetty, where oil and water are loaded before her departure from Grangemouth to return to the Netherlands after almost 20 years (!).

During the return journey to Den Helder, the ship 'accidentally' sails over the Dogger Bank, which had not yet been cleared of mines. On 9 May, 1946, she enters Den Helder. Shortly after, she goes to the Navy yard for repairs and reconstruction. On 15 July, the ship is fit for duty in a new role as a radar training ship.

On 26 July, she arrives at Portsmouth and moors at the Harbour NW Jetty Tridal Basin. After departure *Soemba* conducts radar exercises and proceeds to Sandown Bay where she anchors on 7 August. Over the next two days she executes radar exercises, conducted with an with an aircraft at sea, and then proceeds back to Sandown Bay. After these exercises, she alters course to Spithead (9 - 10 Aug) and Portsmouth (10 - 13 Aug). After departure there is no aircraft available for radar exercises. Instead plotting exercises are now being conducted. In afternoon she arrives at Sandown Bay. The daily routine of exercises at sea and sheltering in port repeats for some days before going to Portsmouth on 16 August.

On 20 August, *Soemba* steams out for radar exercises and then anchors at Spithead Road. The next day she conducts radar exercises with an aircraft, before entering Brighton due to a gale warning. Other ports visited; Portsmouth (23 - 27 Aug) , Brighton (27 - 28 Aug), Spithead (29 Aug) and Portsmouth (29 - 30 Aug). Returning to Rotterdam on 31 August, were *Soemba* moors at Parkkade. On 2 September the Commander of the Netherlands Navy visits the ship. The ship exercises its personnel at sea and tests the radar installation (6 Sep). On 9 September, the radar course for officers begins. Four days later she departs Rotterdam and steams out for a radar course for officers. Returning to Rotterdam (13 -19 Sep) The radar course for officers is held offshore, the ship steams in dueto bad weather.

Soemba ss radar training ship. (National Archives)

Officers and crew of the radar training ship Soemba *in July 1947.*
(National Archives)

the Hook of Holland (19-20 Sept). Due to bad weather, the ship steams to Rotterdam (20 -29 Sept.). After departure, she conducts a radar course for officers at sea. Returning in Rotterdam 27 September on Rotterdam. Until 1 November, *Soemba* makes several of these short training journeys in this period.

On 11 November 1946, the ship's dog gives birth to 9 pups. The 'biological father' of this litter remains unknown.
On 18 November, a new course starts, now for Radio Distance Signallers (officers) and Quartermaster. *Soemba* departs Rotterdam and heads for Portsmouth. The weather conditions on the English south coast are such that it is pointless to exercise outdoors at night. During the day, even for the next period, its nothing more than 'taking shelter' without any oppertunity to conduct exercises, as the radio bearings do not work properly in winds of force 7 to 10. It would take untill 25 November before she departs again for exercises. Again, a period with short journeys follows.

In January 1947, the ship is moored in Rotterdam without any assignment. On 6 May, the commander requests permission to visit the port of Douglas on the Isle of Man during the weekend of 28 June. However, he is informed that, given the precarious situation of the country's foreign exchange, permission cannot be granted. The costs estimated by the commander amount to approximately £150 wages plus food and drinking water.

On 25 August, *Soemba* begins tactical exercises in the North Sea, together with the aircraft carrier *Karel Doorman*.
In December, *Soemba* receives maintenance at the Navy Yard. Once repaired, she stays in port.

In June 1949, the ship sails to Amsterdam. At the Navy Barracks in Amsterdam, the Royal Mention to the Order of the Day (Royal Decree of 10 November 1948, no. 30) is unveiled by Prince Bernhard, in the presence of several high-ranking officers.

On 27 June 1949, *Soemba* exercises with light cruiser *Tromp*, frigate *Van Kinsbergen*, and destroyer *Piet Hein* in the North Sea. The following day, she exercises with minesweeper *Hollandsdiep*, auxiliary *Vulkaan*, and patrol vessel *Queen Wilhelmina*. British and Belgian ships also participate in the exercise. Shortly after, in July, she joins 'Exercise Verity', a Western European naval exercise. in the Bay of Biscay and the English Channel and are commanded by Admiral Rhoderick McGrigor of the British Home Fleet.

On 4 October 1949, *Soemba* is decommissioned.

On 17 June 1947, in the presence of Mrs. I.D.E. Sterkenburg-Dirkzwager a plaque placed on the bridge to commemorate the killing of her spouse and former ship's commander on 55 August 1943.
On 12 August 1983, Mrs. Sterkenburg-Dirkzwager placed the floral tribute for the last time and removed the copper memorial plaque, ending this traditional commemoration ceremony after 36 years due to the ship's planned decommissioning. The copper plaque is nowadays part of the collection in the Navy Museum in Den Helder.

Tour with radar training ship Soemba, *organised for the press and public in 1947.*
(National Archives)

Executing RAS (Replenishment At Sea) in 1949. Receiving oils following the 'heavy weather method' sailing in the wake of the supply ship, who streams a hose in the water. Soemba *has already retrieved and connected the hose.* (NIMH)

By Royal Decree No. 30 of 10 November 1948, the gunboat Soemba was awarded a Royal Mention in the Order of the Day.

At the Navy Barracks in Amsterdam, the Royal Mention to the Order of the Day is unveiled by Prince Bernhard, in the presence of Vice Admiral E.J.v.Holthe, Chief of Naval Staff and BDZ, and Rear Admiral J.J.L.Willinge, Chief of Naval Staff of the Netherland.

WIJ JULIANA, BIJ DE GRATIE GODS KONINGIN DER NEDERLANDEN, PRINSES VAN ORANJE-NASSAU, ENZ., ENZ.,

Op de voordracht van Onze Minister van Marine a.i. van 29 October 1948, nr. 150172/13046;

HEBBEN GOED GEVONDEN EN VERSTAAN:

Toe te kennen de „Koninklijke Vermelding bij Dagorder" aan Onze kanonneerboot „Soemba".

„Onze Kanonneerboot „Soemba" heeft zich in de Aziatische wateren onderscheiden van December 1941 tot half Maart 1942 door uitstekende daden bedreven door de bemanning tegenover de vijand bij het uitvoeren van haar opdrachten in het zeegebied van Indonesië, mede tijdens vijandelijke vliegtuigaanvallen;

heeft zich daarna in de Europese wateren wederom onderscheiden: door de uitstekende daden van de bemanning, daarbij met veel succes en tot grote voldoening van de geallieerde bevelvoering directe daadwerkelijke artilleriesteun te verlenen bij de landingsoperaties op de Oostkust van Sicilië in Juli en Augustus 1943, op het Anzio- en Formiafront in Januari en Februari 1944 en uiteindelijk op de kust van Normandië in Juni en Juli 1944, bij welke operaties zware vijandelijke tegenstand — ook uit de lucht — werd ondervonden en op 19 Juli 1944 een vijandelijk vliegtuig werd neergeschoten;

bovendien door de eer van de Nederlandse vlag op de vrije wereldzee hoog te houden door in het tijdvak van 21 April tot 4 December onafgebroken te patrouilleren in de Perzische Golf ter beveiliging van het voor de oorlogvoering zo uiterst belangrijke olievervoer en het gedurende de gehele wereldoorlog 1940—1945 telkermale escorteren van tal van belangrijke scheepsconvooien in de Indische Oceaan, de Middellandse Zee en de Atlantische Oceaan".

Soestdijk, 10 November 1948

w.g. JULIANA.

1950's

At the end of 1949, an extensive conversion into an aircraft control ship started. During the ship's conversion, parts and equipment are used from the British battleship *Rodney*, which was scrapped in 1948. By mid-1951 she is again fit for duty. She is now used for the training of radio distance finders to familiarise naval personnel who were not in England during the war with the latest developments in this field. *Soemba* makes several training trips, including to England.

On 18 June 1951, she participates in the Allied naval exercise 'Castanets', in which asquadron of 12 minesweepers, and the frigate *Marnix* are also present. Also, *Soemba* exercises with the Minesweeper Flotilla 4 and participates in the NATO exercise 'Factotum'.

On 4 March 1952, the ship undergoes the necessary repairs at the Navy Yard.
In early May 1953, she exercises near Plymouth with the British radar training ship HMS *Boxer*. Afterwards, the ship undergoes maintenance at the Rijkswerf. During the maintenance period, the ship visits Amsterdam for the Naval Review in June. In August, maintenance is completed. On 19 September, *Soemba* participates in the major NATO exercise 'Mariner' and. on 26 November, she exercises again near Portsmouth with HMS *Boxer*, before undergoing maintenance at the Navy Yard.

In April 1954, *Soemba* participates in exercises in the North Sea. On 6 April, she rendezvouses with Task Group 5 and joins in the final exercises of the Task Group, which are on their homeward voyage. *Soemba* arrives at Sandown Bay the next day. More exercises are conducted from this location.

The 'work floor' of radar direction vessel Soemba. *(National Archives)*

Leaving the Task Group on 13 April and heads for Portsmouth. On 11 May 1954, *Soemba* exercises again in the Channel with HMS *Boxer*.After some weeks. she participates in the Fleet Week in Rotterdam (12 - 17 Jun) and, on 14 July in NATO Exercise "Haul". This exercise is intended to practice the execution of the wartime tasks of the naval forces belonging to the Channel Command. In addition to Task Group 5. In October the ship has to go

Left:
Soemba, *moored aft of frigate* Marnix *(F 801) former HMS* Garland *(H 38), at the Fleet Week in Rotterdam 1954.* *(National Archives)*

To provide an earlier warning for attacks. Soemba *was stationed beyond the fleets radar horizon. To protect herself she was fitted with AA-guns and some comprehensive radar warning and jamming devices of her day.* (NIMH)

to the Navy Yard because of the poor condition of her boilers. After a thorough inspection it is concluded that repair is not cost-effective and on 30 November, 1954 she is decommissioned pending removal from service.

On 1 January, 1956, *Soemba* was removed from service. It is decided to convert the ship to an accommodation vessel, for which purpose it is converted at the Navy Yard.

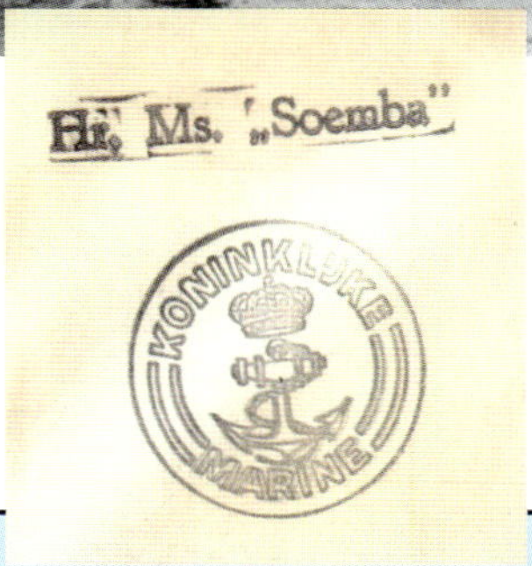

Postal stamp 1950s

Twin sister

The *Flores*, still possessed usable 15cm guns and remained in service as a gunboat, returning to the Netherlands in 1946. She was assigned various tasks, such as ferry services, sailing training personnel to London, acting as an artillery training ship, and providing technical training. In 1950, the gunboat was classified as a frigate but primarily served as an accommodation ship. In 1955, the ship was decommissioned and given a permanent berth in Vlissingen as an accommodation ship. On August 26, 1968, *Flores* was definitively decommissioned and sold for scrapping at Hendrik-Ido-Ambacht.

Accomodation vessels Flores *(A 877), still with guns, and* Jacob van Heemskerck *(A 879) in 1955.* (Collection T.L. Sander)

Accomodation vessels Flores *(A 877) and* Willem van der Zaan *(A 880) in the outer harbour of Vlissingen, 1965.*

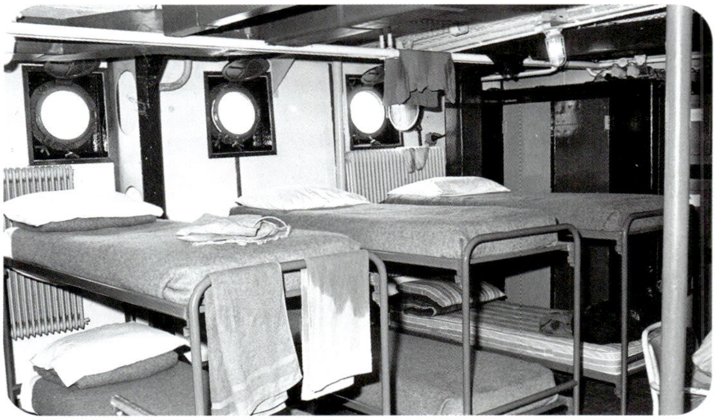

Since 1 November 1956, *Soemba* is placed at the disposal of the Head of the Conservation Service in Den Helder and intended as an accommodation ship for the Mine warfare Department for the training of divers in Den Oever. Beforehand, *Soemba* is transferred from Den Helder to Den Oever on 15 September 1958 to relieve *Castor* as accommodation ship for the diving company. After 4 years, *Soemba* was reconstructed at the navy yard for an accommodation expansion. The work lasted from December 1962 to 1964. The former minelayer *Medusa*, replaced her.

Accommodation on board. (NIMH)

Besides accommodation ship for the Diver training, the Amphibious Section of the Marine Corps is also stationed aboard the ship.

For many years a familiar sight; Soemba and dive tender Roermond in harbour of Den Oever. (NIMH)

On 1 March 1965, the mooring pontoon *Y 8711* is added to *Soemba*.

On 13 December 1968, a new shipscrest was allocated to *Soemba*. As this was appopriate to her contemporary role.
By 1969, 38 courses are given on board in the fields of diving, dismantling, demolition, as well as 13 courses on knowledge of diver's illnesses. A total of 186 students are trained. *Soemba* has two diving sloops at its disposal for training: the *Triton* and the *Nautilus*, and for disassembly training on the diving simulator *Y 8690*, which was acquired from the navy yard on 16 December.

The former harbour tug *Y 8015* was sunk in the IJsselmeer on 9 July 1969, to serve as a practice wreck.
A year later, in 1970, thirty-six courses were given on board in the fields of diving, disassembly, demolition, as well as 23 courses on diver's illness. A total of 210 students were trained. For training purposes, *Soemba* has two diving sloops, a landing craft, a mine-sweeping service boat, the pontoon *Y 8711*, and the *Y 8690* for disassembly training.

On 12 February 1975, a lieutenant 2nd class Technical Service went missing during a diving exercise in the North Sea north of Den Helder. Later the remains are recovered.

Above: *Accomodation ship* Soemba. (NIMH)

The ship in 1984. Photographs by C. Sander. (Collection T.L. Sander)

On her way to the ship breaker in Belgium. (NIMH)

References

- Door de jaren heen Hr.Ms. kanonneerboot *Soemba*, Cruijff, Henk, newsletter Navy Museum, Den Helder October 2010.
- Hr.Ms. Flottieljevaartuigen *Flores* en *Soemba*, *Het Schip*, 8e jaargang No.17, August 1926
- Kanonneerboten Floris en Soemba, Olling, C.L.H., *Uit de Miljoenenhoek*, 78 jaargang No.5, Mei 1989
- *Naval Radar*, Friedman, Norman, Naval Institute Press 1980
- *Radar Development in The Netherlands, 100 years since Hülsmeyer*, Piet van Genderen ea.,
- *Emblemen van de Koninklijke Marine*, Eekhout, L.L.M., Schutte, O., and Pol, P.J.F. van der, 1991 Eisma B.V.
- *USNI Historical Atlas of U.S. Navy*, Graig L. Symonds, 1995 Naval Institute Press.

- *Jaarboek Koninklijke Marine* (various volumes)
- *Warship World* (various issues)
- Website: Dutchfleet.nl

On 14 June 1976, the *Y 8512* (nicknamed "*Augurk*") is added to *Soemba*. Still acting as an instruction vessel for the dive courses like searching for missing people in the water, placing listening devices on boats, and searching for mines underwater.
The copper memorial plaque that had been mounted on the bridge 36 years ago was removed on 12 August 1983 and offered to the Navy Museum.
The general condition of *Soemba* had gradually deteriorated. For these reasons, the training and courses on board were stopped on 28 January 1985. The ship was then prepared to be towed to Den Helder, which became a reality on 14 March after which the final decommissioning took place with a ceremony on 7 June 1985. Rejected on 2 June 1985.Although initiatives arose to save her, she was sold for scrap about 4 weeks later. On 11 July, she was towed away by *Taktow 1* to Antwerp.

Series editor
Jantinus Mulder

Publisher
Walburg Pers/Lanasta

Authors
Jantinus Mulder

Contributors
Dick Vries

Graphic design
Jantinus Mulder

Translation revision
Deben Translations

First print, July 2026
ISBN 978-94-6456-695-6
NUR 465

Contact Warship
jantinusmulder@walburgpers.nl

Lanasta

www.walburgpers.nl/lanasta

Wachten op het einde...

Behalve de onderzeeboot Dolfijn valt momenteel nòg een oud-marineschip onder de slopershamer. Het is de Soemba, tijdens de laatste wereldoorlog een roemruchte kanonneerboot en voor het laatst dienst gedaan hebbend als logement- en instructievaartuig voor marineduikers in Den Oever. De Soemba werd in juli uit Den Helder weggesleept, in opdracht van Heuvelman Staal BV uit 's Gravendeel. Het bedrijf was voor 75.000 gulden eigenaar geworden, maar heeft het schip meteen doorverkocht naar een scheepssloperij in het Belgische Burcht, een gehucht westelijk van Antwerpen. De Soemba werd daar gedumpt op een droogvallend stuk van de Schelde, waar zij meteen met schroeven en roer in de blubber wegzakte. Wachtend op het einde...

A small item in a newspaper. The headline reads: Waiting for the end...
After arrival at the shipbreaker in Burcht (B), Soemba *was dumped there on a drying stretch of the Scheldt, where she almost instantly sank into the mud, propellers and rudder intact.*